T0192088

Universitext

Springer

New York
Berlin
Heidelberg
Barcelona
Budapest
Hong Kong
London
Milan
Paris
Santa Clara
Singapore
Tokyo

Universitext

Editors (North America): S. Axler, F.W. Gehring, and K.A. Ribet

Aksoy/Khamsi: Nonstandard Methods in Fixed Point Theory
Andersson: Topics in Complex Analysis
Aupetit: A Primer on Spectral Theory
Booss/Bleecker: Topology and Analysis
Borkar: Probability Theory: An Advanced Course
Carleson/Gamelin: Complex Dynamics
Cecil: Lie Sphere Geometry: With Applications to Submanifolds
Chae: Lebesgue Integration (2nd ed.)
Charlap: Bieberbach Groups and Flat Manifolds
Chern: Complex Manifolds Without Potential Theory
Cohn: A Classical Invitation to Algebraic Numbers and Class Fields
Curtis: Abstract Linear Algebra
Curtis: Matrix Groups
DiBenedetto: Degenerate Parabolic Equations
Dimca: Singularities and Topology of Hypersurfaces
Edwards: A Formal Background to Mathematics I a/b
Edwards: A Formal Background to Mathematics II a/b
Foulds: Graph Theory Applications
Friedman: Algebraic Surfaces and Holomorphic Vector Bundles
Fuhrmann: A Polynomial Approach to Linear Algebra
Gardiner: A First Course in Group Theory
Gårding/Tambour: Algebra for Computer Science
Goldblatt: Orthogonality and Spacetime Geometry
Gustafson/Rao: Numerical Range: The Field of Values of Linear Operators
    and Matrices
Hahn: Quadratic Algebras, Clifford Algebras, and Arithmetic Witt Groups
Holmgren: A First Course in Discrete Dynamical Systems
Howe/Tan: Non-Abelian Harmonic Analysis: Applications of $SL(2, R)$
Howes: Modern Analysis and Topology
Humi/Miller: Second Course in Ordinary Differential Equations
Hurwitz/Kritikos: Lectures on Number Theory
Jennings: Modern Geometry with Applications
Jones/Morris/Pearson: Abstract Algebra and Famous Impossibilities
Kannan/Krueger: Advanced Analysis
Kelly/Matthews: The Non-Euclidean Hyperbolic Plane
Kostrikin: Introduction to Algebra
Luecking/Rubel: Complex Analysis: A Functional Analysis Approach
MacLane/Moerdijk: Sheaves in Geometry and Logic
Marcus: Number Fields
McCarthy: Introduction to Arithmetical Functions
Meyer: Essential Mathematics for Applied Fields
Mines/Richman/Ruitenburg: A Course in Constructive Algebra
Moise: Introductory Problems Course in Analysis and Topology
Morris: Introduction to Game Theory
Polster: A Geometrical Picture Book
Porter/Woods: Extensions and Absolutes of Hausdorff Spaces
Ramsay/Richtmyer: Introduction to Hyperbolic Geometry
Reisel: Elementary Theory of Metric Spaces
Rickart: Natural Function Algebras

(continued after index)

L.R. Foulds

Graph Theory Applications

With 90 Illustrations

 Springer

L. R. Foulds
Department of Management Systems
University of Waikato
Hamilton, New Zealand

Mathematics Subject Classification (1991): 05-01, 49-10, 90-01

Section 7.6 is based on material from Chapter 4 of *Digraphs: Theory and Techniques*, by Robinson and Foulds and published by Gordon and Breach in 1980. This material is included here with the permission of Gordon and Breach.

Library of Congress Cataloging-in-Publication Data
Foulds, L.R., 1948–
      Graph theory applications / L.R. Foulds.
         p.   cm. — (Universitext)
      Includes bibliographical references and index
      ISBN 0-387-97599-3 (acid-free paper)
      1. Graph theory.   I. Title.   II. Series.
QA166.F68   1991
511´.5—dc20   91-20590

Printed on acid-free paper.

Photocomposed copy prepared using the author's TEX file.
Printed and bound by Edwards Brothers, Inc., Ann Arbor, MI.
Printed in the United States of America.

9 8 7 6 5 4 3

ISBN 0-387-97599-3 Springer-Verlag New York Berlin Heidelberg
ISBN 3-540-97599-3 Springer-Verlag Berlin Heidelberg New York   SPIN 10677249

Preface

Over the last 30 years graph theory has evolved into an important mathematical tool in the solution of a wide variety of problems in many areas of society. The purpose of this book is to present selected topics from this theory that have been found useful and to point out various applications. Some important theoretical topics have been omitted as they are not essential for the applications in Part II. Hence Part I should not be seen as a well-rounded treatise on the theory of graphs. Some effort has been made to present new applications that do not use merely the notation and terminology of graphs but do actually implement some mathematical results from graph theory. It has been written for final undergraduate year or first year graduate students in engineering, mathematics, computer science, and operations research, as well as researchers and practitioners with an interest in graph theoretic modelling. Suggested plans for the reading of the book by people with these interests are given later. The book comprises two parts. The first is a brief introduction to the mathematical theory of graphs. The second is a discussion on the applications of this material to some areas in the subjects previously mentioned. It is, of course, possible to read only the first part to attempt to gain an appreciation of the mathematical aspects of graph theory. However even the purest of mathematicians is strongly recommended to delve seriously into the second part. This is because the theory of graphs and the applications of graphs are inextricably intertwined. Much of the mathematical theory of graphs has arisen out of attempts to solve practical problems. So to ignore the utility of graph theory is to ignore a major part of its importance.

The text evolved out of the experience of the author in teaching the material to students in mathematics at Massey University, operations research at the University of Canterbury, operations management at the University of Waikato, (all in New Zealand) engineering at the University of Florida, and management information systems at University College Dublin. It contains exercises which the reader is urged to try. As with all disciplines, you cannot master graph theory without getting your hands dirty. Graph theory is not a spectator sport!

The first 10 chapters, making up the first part, are organised as follows. We begin with an introductory chapter which introduces a little historical background and the fundamental notions. It is assumed in all later chapters that the reader is familiar with this material. Chapter 2 is concerned with connectivity — a concept which is basic for many of the later chapters. One of the most important classes of graphs is that of trees and this is dealt with in Chapter 3. The material in the next chapter, on traversability, is of theoretical interest in its own right but is also of practical importance in operations research, covered in Chapter 12. Chapter 5, on planarity, illustrates that the topic is more than just topology. It is necessary for the sections on layout in Chapter 14. Chapter 6, on the matrices of a graph, is essential for a later discussion, on graph theoretic algorithms. Chapter 7 is a necessarily brief account of directed graphs, called digraphs, and their important special case; the network. Chapter 8, on covering, dominance, and matching has application in industrial engineering and other disciplines. Chapter 9 covers graph theoretic algorithms. In Chapter 10 we make a brief excursion into the world of matroids, where there are applications in electrical engineering, among other areas. Part II has mainly longer chapters explaining the application of the above-mentioned material in various branches of engineering, operations research, and science. No attempt has been made to make this part encyclopaedic. Rather, due to limititations of space and for other reasons, just a few applications have been presented in some depth. They are intended to give some impression of the power and wide utility of graph theory.

Part I is suitable as a one-semester course in mathematics or engineering and this could be followed by a second semester covering the applications in Part II. Other one-semester sequences are given later.

A few of the chapter sections and exercises are starred. These require a greater level of mathematical maturity. They may be skipped without loss of continuity.

The author would like to thank the University of Waikato, and University College Dublin. The former institution allowed the author to write this book in the course of his employment. The latter institution hosted the author while he was on sabbatical leave during the 1988–89 year, (along with Trinity College, Dublin) during which time the book was tested on its students.

The author is also grateful to Nikki Sayer who typed the entire manuscript, to Erica Harris for wordprocessing the manuscript using TeX, to Frank Bailey and Doreen Whitehead of the Draughting Department at the University of Waikato for drawing the figures, to Derek O'Connor and Fergus Gaines of University College Dublin, Takeo Yamada and his students at the National Defense Academy, Yokosuka, Japan, Horst Hamacher of the University of Kaiserslautern, Germany, and to Professor J. Labelle of the University of Montreal, who all read the entire manuscript and who suggested many valuable improvements, and to the staff of Springer-Verlag for their able cooperation.

L.R. Foulds
Hamilton, New Zealand

To Jacqueline Beaton
The LFJB

Contents

Preface   v

Teaching Plans   xv

Part I: The Theory Of Graphs   1

Chapter 1:   BASIC IDEAS   3
                 History   3
                 Initial Concepts   9
                 Summary   15
                 Exercises   15

Chapter 2:   CONNECTIVITY   17
                 Introduction   17
                 Elementary Results   19
                 Structure Based on Connectivity   22
                 Summary   24
                 Exercises   25

Chapter 3:   TREES   27
                 Characterizations   27
                 Theorems on Trees   30
                 Tree Distances   30
                 Binary Trees   32
                 Tree Enumeration   35
                 Spanning Trees   37
                 Fundamental Cycles   38
                 Summary   39
                 Exercises   41

Chapter 4:   TRAVERSABILITY   43
                 Introduction   43
                 Eulerian Graphs   43
                 Hamiltonian Graphs   46
                 Summary   50
                 Exercises   51

Chapter 5:   PLANARITY   53
                 The Utilities Problem   53
                 Plane and Planar Graphs   55
                 Planar Graph Representation   55
                 Planarity Detection   60
                 Duality   64
                 Thickness and Crossing Numbers   70
                 Summary   73
                 Exercises   73

Chapter 6:   MATRICES   75
                 The Adjacency Matrix   76
                 The Incidence Matrix   77
                 The Cycle Matrix   80
                 The Cut-Set Matrix   84
                 The Path Matrix   90
                 Summary   91
                 Exercises   91

Chapter 7:   DIGRAPHS   93

Connectivity   93

Traversability   98

Directed Trees   100

More Digraph Matrices   100

The Principle of Directional Duality   107

Tournaments   108

Summary   120

Exercises   120

Chapter 8:   COVERINGS AND COLOURINGS   123

Covering, Independence, and Domination   124

Colouring   132

Matching   134

Summary   142

Exercises   143

Chapter 9:   ALGORITHMS   145

Algorithms   146

Input   146

Complexity   149

Output   160

Graph Analysis Algorithms   161

Graph Optimization Algorithms   174

Summary   180

Exercises   180

Chapter 10:   MATROIDS   183

Introduction   184

Duality   186

The Greedy Algorithm   188

Summary   191

Exercises   191

Part II: Applications   193

Chapter 11:   MISCELLANEOUS APPLICATIONS   195
                    Social Sciences   197
                    Economics   199
                    Geography   202
                    Architecture   207
                    Puzzles and Games   210
                    Summary   220
                    Exercises   221

Chapter 12:   OPERATIONS RESEARCH   225
                    Operations Research and Graph Theory 226
                    Graph Theoretic Algorithms in OR   226
                    Graph Theoretic Heuristics in OR   231
                    Digraphs in OR   233
                    Optimization Algorithms   234
                    Transportation Networks: Advanced Models 261
                    Summary   265
                    Exercises   265

Chapter 13:   ELECTRICAL ENGINEERING   269
                    Electrical Network Analysis   269
                    Printed Circuit Design   279
                    Summary   289
                    Exercises   289

Chapter 14:   INDUSTRIAL ENGINEERING   291
                    Production Planning and Control   291
                    Facilities Layout   292
                    Summary   318
                    Exercises   319

Chapter 15: SCIENCE 323

Physics 323

Chemistry 324

Biology 328

Summary 340

Exercises 340

Chapter 16: CIVIL ENGINEERING 343

Earthwork projects 343

Traffic Network Design 344

Summary 358

Exercises 358

Further Reading 361

Bibliography 365

Index 379

Part II: Applications   193

Chapter 11: MISCELLANEOUS APPLICATIONS   195
                 Social Sciences   197
                 Economics   199
                 Geography   203
                 Architecture   207
                 Puzzles and Games   210
                 Summary   220
                 Exercises   221

Chapter 12: OPERATIONS RESEARCH   225
                 Operations Research and Graph Theory   226
                 Graph Theoretic Algorithms in OR   226
                 Graph Theoretic Heuristics in OR   231
                 Digraphs in OR   233
                 Optimization Algorithms   234
                 Transportation Networks: Advanced Models 261
                 Summary   265
                 Exercises   265

Chapter 13: ELECTRICAL ENGINEERING   269
                 Electrical Network Analysis   269
                 Printed Circuit Design   279
                 Summary   289
                 Exercises   289

Chapter 14: INDUSTRIAL ENGINEERING   291
                 Production Planning and Control   291
                 Facilities Layout   292
                 Summary   319
                 Exercises   320

Teaching Plans

The book is suitable as a course in a number of disciplines. Here are some chapter sequences:

Mathematics:

1, 2, 3, 4, 5, 6, 7, 8, 9, 10, 11.

Computer Science:

1, 2, 3, 4, 5, 6, 7, 8, 9, 11.

Management Science or Operations Research:

1, 2, 3, 4, 5, 6, 7, 8, 9, 11, 12.

Industrial or Systems Engineering:

1, 2, 3, 4, 5, 6, 7, 8, 9, 14.

Urban Planning or Civil Engineering:

1, 2, 3, 4, 5, 6, 7, 8, 9, 16.

Electrical Engineering:

1, 2, 3, 4, 5, 6, 7, 8, 9, 13.

Science:

1, 2, 3, 4, 5, 6, 7, 8, 9, 15.

Graph Theory Applications

Part I: The Theory Of Graphs

"It is a capital mistake to theorize before one has data."

A Scandal in Bohemia, (Sherlock Holmes) Arthur Conan Doyle

The purpose of Part I of this book is to present selected topics from the theory of graphs, the choice being guided by what is likely to be applicable for the engineer, scientist, or worker in operations research. Naturally, when discussing a field which is likely to be new to most of the intended readers, we must introduce the basic concepts of the subject and the notation and terminology which describes them. In doing this we attempt to motivate why the concepts of graph theory came into being and to hint at why they are important in today's technological world.

1

1 Basic Ideas

"Mighty things from small beginnings grow".

Annus Mirabilis, John Donne

In this first chapter we introduce the subject of graph theory by first presenting some of the problems, both frivolous and practical, that led mathematicians to discover and rediscover some of the fundamental concepts of the subject. After a brief historical introduction, we set the theory of graphs on a firm mathematical foundation with some basic concepts and definitions.

1.1 History

1.1.1 The Königsberg Bridge Problem

On Sunday afternoons in the eighteenth century the good citizens of the city of Königsberg in Eastern Prussia, would amuse themselves with the following puzzle. Through the city flowed the lovely River Pregel, passing around the island of Kneiphof, as can be seen in Figure 1.1. The puzzle involved starting on any one of the pieces of land and finding a tour that crossed each bridge exactly once, returning to the starting point. Since none of the populace could achieve this feat, it came as no surprise when the great Swiss mathematician L. Euler proved it to be impossible. Euler (1736)

further, gave precise conditions for when such a tour could be found for an
arbitrary system of interconnected bridges.

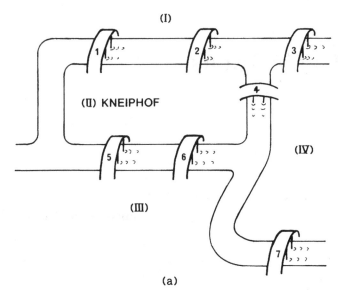

(a)

Figure 1.1 Eighteenth Century Königsberg

The nature of Euler's paper led him to be called *the father of graph theory*.
This is because the land areas and the bridges in Figure 1.1 can be represented by an equivalent system of points and lines as shown in Figure 1.2,
thus producing a picture of a *graph*. By analyzing Figure 1.2, we can see
that the problem reduces to finding a sequence of points and lines which
traverses each line exactly once, and which returns to its starting point.
Euler proved that this is possible only if every point is incident with an
even number of lines. This condition certainly does not hold for the system
in Figure 1.2, where none of the points obeys it. Hence the desired tour
cannot be found. These ideas are discussed in more depth in Section 4.2.

1.1.2 Hamilton's Game

In 1856 the Astronomer Royal of Ireland, W.R. Hamilton, presented to the
world the following puzzle. On the diagram in Figure 1.3, which can be
seen to be the graph of the dodecahedron, one must find a tour passing

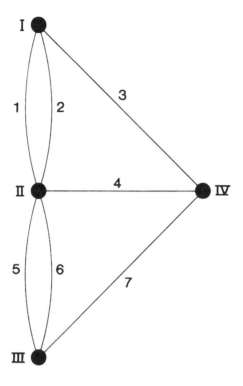

Figure 1.2 The graph of the Königsberg
bridge problem

along the indicated lines, visiting each point exactly once, and returning to a given starting point.

To make things a little more glamorous, a later version of the game was manufactured using the names of various cities to label the points, e.g. Brussels, Canton, Delhi, ... and so on. Hence one of the players was asked to find *a tour around the world*. Hamilton sold the game to a London games maker for 25 pounds, much to the later chagrin of the latter, as the game was not a commercial success.

In viewing the skeleton of the dodecahedron as a graph, we can see that the original puzzle asks a question similar to that of the Königsberg bridge problem: Find a tour in the graph with a certain given property. In the

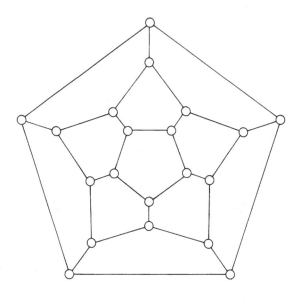

Figure 1.3   Hamilton's game; the graph of the
dodecahedron

previous section the property was: Pass along each line exactly once. In
the present section the property is: Pass through each point exactly once.
Surprisingly, the two problems are not similar from a mathematical point
of view. The former has quite a simple characterization. No reasonable
characterizations are known for the latter problem. However we shall study
what is known in Section 4.3.

1.1.3 The Four Colour Theorem

Until 1976, proving the following conjecture was one of the most famous
unsolved problems in mathematics:

> **Any map on a plane surface (or a sphere) can
> be coloured with at most four colours so that no
> two adjacent regions have the same colour.**

The conjecture can be posed in graph theoretic terms as follows. Represent
the regions of the map by points and join neighbouring pairs by lines. Then

the problem is to colour the points of the graph so that no two are joined by a line having the same colour. The problem seems to have surfaced when Francis Guthrie (later to become professor of mathematics at Cape Town) posed it to his brother Frederick, a student at University College, London. Frederick Guthrie (1880) mentioned it to his mathematics professor, Augustus De Morgan, who wrote of it to W.R. Hamilton. Although Hamilton took no interest in the conjecture, De Morgan probably communicated the problem to the English mathematician Arthur Cayley (who made it known to the Royal Geographical Society) and to a London lawyer A.B. Kempe (1879) who published a false "proof", appearing to settle the conjecture in the affirmative. P.J. Heawood (1890) found a flaw in Kempe's proof. K. Appel and W. Haken (1976), showed that any map must contain at least one of a certain set of 1936 map configurations. If the map cannot be coloured in four colours, because it has one of the special map configurations, there must be a smaller map which cannot be coloured in four colours. But if the original map is assumed to be minimal in the sense that any of its submaps can be coloured in four colours, a contradiction ensues. Haken and Appel used these ideas to settle the conjecture in the affirmative.

These, and other graph colouring matters, are discussed in greater depth in Section 8.2.

1.1.4 Graphs in Chemistry

The term *graph*, as the name of a system of points and lines, came from the phrase *graphic notation*, first introduced in chemistry by E. Frankland and adopted, in 1884, by A. Crum Brown (1884). Each atom of a chemical structure is represented by the point of a graph and atomic bonds are represented by lines, as shown in Figure 1.4. The two graphs in the figure both have the chemical formula C_3H_7OH, but are different molecules and have significantly different physical properties. Chemical compounds exhibiting this phenomenon are called *isomers*. The mathematician A. Cayley, mentioned earlier, noticed that many of these graph theoretic representations have the property that there is a unique path of lines between each pair of points. A graph with this property is called a *tree*. Indeed, trees are among the most important of graphs. Cayley earlier introduced the concept of a tree when observing changes in variables in the calculus. He then attempted to use this work to enumerate the possibilities for certain isomers which are trees, including the paraffins, C_kH_{k+2}. He also managed to enumerate all rooted trees, where one point is specially labelled (Cayley (1857)) and trees where every point is incident with at most four lines. These, and other tree enumeration problems, are discussed in greater depth in Section 3.5. Some applications of graph theory in chemistry are outlined in Section 15.2.

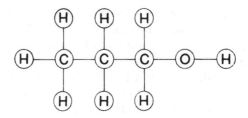

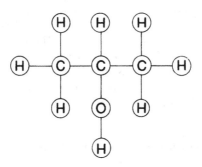

Figure 1.4 Chemical isomers

1.1.5 Graphs in Physics

In 1845, G.R. Kirchhoff announced two rules that are thought to govern the flow of electric current in a network of wires. One of these is that the potential difference around any circuit in the network sums to zero. The other is that the algebraic sum of the current flowing into the network junction is zero. It is clear that we can abstract a network as a graph, with junctions and wires being represented by points and lines, respectively. Kirchhoff (1847) stated that it is not necessary to examine every circuit in order to deduce the current flow in each wire of the network. Indeed, he answered the natural question as to how many circuits, and which ones, must be analyzed in order to obtain the complete picture. A set which leads to the desired result is called a *set of fundamental circuits*. Kirchhoff's method for identifying such a set, which is still in use today, is as follows.

Begin by finding a subset of the lines of the underlying graph which forms a tree (in the sense defined in the previous section) and which contains every point of the network. The addition of any other line of the graph produces exactly one circuit. Each such circuit, produced by the addition of exactly one line to the tree, is a fundamental circuit. Thus there are as many fundamental circuits as there are additional lines, over and above the lines in the initial tree. Of course the fundamental set produced by this method is dependent upon the original choice of the tree. These ideas are elaborated upon, when we discuss the application of graph theory in electrical engineering, in Section 13.1.

1.2 Initial Concepts

Since the work by L. Euler (see Section 1.1.1), there has been a wide divergence in graph theoretic terminology. This is not altogether a bad thing as it allows for the efficient use of terms which are selected with certain purposes in mind. The terminology presented here is a selection of what is considered by many mathematicians to be in common use.

Unfortunately there is a formidable number of definitions which must be presented, so we ask the reader to have patience — the tedium of the rest of this chapter will make the heady wine to come all the more enjoyable. A *graph* (V, E), is an ordered pair, where V is a finite, nonempty set whose elements are termed *vertices*, and where E is a set of unordered pairs of distinct vertices of V. Each element $\{p, q\} \in E$ (where $p, q \in V$), is called an *edge* and is said to *join* the vertices p and q. (Some authors allow V to be possibly empty or infinite. However most graph theoretic applications involve a finite (but nonempty) set of vertices and thus we make no such allowances here.)

Synonyms for *vertex* are: *point, node,* and *junction.* Synonyms for *edge* are: *line, arc, branch,* and *link.* If the edge $e = \{p, q\} \in E$, then p and q are both said to be *incident with* e and *adjacent to* each other. Further, if edge $e' = \{p, r\} \in E$, then the edges e and e' are said to be *adjacent* as they have a vertex (in this case p) in common. We denote the vertices with which a vertex v, is adjacent by $\Gamma(v)$. For brevity, an edge $\{p, q\}$ is usually denoted by pq.

One of the features which makes graph theory fun is the opportunity to represent any graph by a picture. This is always possible because V (and hence E) is finite. A pictorial representation can be achieved by representing each vertex in V by a point in the plane and each edge in E by a nonintersecting line joining the points representing the vertices with which it is incident. As an example, the graph $G = (V, E)$, where $V = \{v_1, v_2, v_3, v_4\}$, and $E = \{v_1v_2, v_2v_3, v_3v_4, v_4v_1, v_2v_4\}$, is shown in Figure 1.5(a). Pictures of other graphs are given throughout this book.

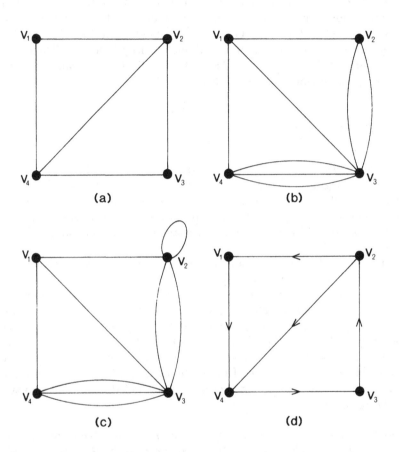

Figure 1.5 (a) Graph, (b) Multigraph, (c) Pseudograph
(d) Digraph

There are many useful structures which bear close resemblance to the structure of a graph. Recall that no more than one edge can join any pair of vertices. Structures in which only this constraint is relaxed are called *multigraphs*. Thus a *multigraph* is an ordered pair of sets (V, E), where V is finite and nonempty, and E is a class of unordered pairs of distinct elements of V where repetitions in the class are allowed. An example of a multigraph is given in Figure 1.5(b). Recall that each edge in a graph and in a multi-graph joins two *distinct* vertices. Structures which, but for this constraint, would be multigraphs are called *pseudographs*. Thus a *pseudograph* is an ordered pair of sets (V, E), where V is finite and nonempty and E is a class of unordered pairs of elements of E where repetitions in the class are allowed. An example of a pseudograph is given in Figure 1.5(c).

Recall that each edge in a graph (V, E), is defined by an *unordered* pair of vertices of V. Sometimes it is useful to give each edge an orientation, or direction. When this has been done the graph is termed a *directed graph* or *digraph* for short. A *digraph* is defined to be an ordered pair of sets (V, A), where V is finite and nonempty and A is a set of ordered pairs of distinct elements of V. We call the elements of A *arcs*, otherwise the terminology for digraphs is similar to that of graphs, given earlier. An example of a digraph is given in Figure 1.5(d). If (u, v) (denoted by uv for brevity) is an arc of a digraph, u is said to be a *predecessor* of v, and v is said to be a *successor* of u.

A graph $G = (V, E)$, or a digraph $D = (V, A)$, is termed *weighted* if there exists a function $w : E \rightarrow \Re$, or $w : A \rightarrow \Re$ (where $\Re$ is the set of real numbers), which assigns a real number, called a *weight*, to each edge of E, or arc of A. Each weight $w(uv)$ where $uv \in E$ or $uv \in A$, is usually written w_{uv}. There is a special class of digraphs called *networks* which we shall study in some depth in Chapters 12 and 16.

We now introduce some concepts used in the definition of a network. If $a = (v_i, v_j)$, is an arc in a digraph (V, A), then a is said to be *directed away* from v_i and *directed towards* v_j. Any vertex which has no arcs directed towards it is said to be a *source*. Any vertex which has no arcs directed away from it is said to be a *sink*. (Of course in some digraphs there are vertices which are neither sources nor sinks.) A *network* is a digraph which possesses exactly one source and exactly one sink. We call the vertices of a network, *nodes*. In most applications of networks there is at least one commodity flowing from the source to the sink in the system which the network represents. An example of a network is given in Figure 1.5(e).

In a similar manner to graphs, it is possible to relax the constraint in the definition of a digraph that no more than one arc can join any pair of vertices. The resulting structures are called *multidigraphs*. Thus a *multidigraph* is an ordered pair of sets (V, A), where V is finite and nonempty and A is a class of ordered pairs of distinct elements of V where repetitions in the class are allowed. An example of a multidigraph is given in Figure 1.5(f). As with graphs, we have required that each arc in a digraph, and in a multidigraph, joins two *distinct* vertices. In the spirit of our discussion on graphs, structures in which this constraint is relaxed are called *pseudodigraphs*. Thus a *pseudodigraph* is an ordered pair of sets (V, A), where V is finite and nonempty and A is a class of unordered pairs of el-

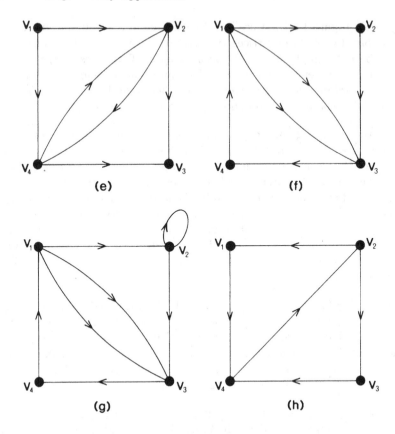

Figure 1.5   (e) Network (f) Multidigraph (g) Pseudodigraph
(h) Oriented graph

ements of A where repetitions in the class are allowed. An example of a pseudodigraph is given in Figure 1.5(g). An arc of the form (v, v), $v \in V$ is termed a *loop*.

It can be seen from the definition that a digraph can have an oppositely directed pair of arcs joining the same pair of vertices. The pair $v_2 v_4$ and $v_4 v_2$ in Figure 1.5(e) provides an example. When a digraph (V, A), possesses at most one of the pair of arcs $v_i v_j$ and $v_j v_i$, for every pair of vertices $v_i, v_j \in V$, it is termed an *oriented graph*. An example of an oriented graph is given in Figure 1.5(h).

A graph (V, E), is termed *labelled* when its vertices in V are distinguishable one from another because they have been named. Thus the graph in Figure 1.5(a) is labelled.

Two graphs $G_1 = (V_1, E_1)$ and $G_2 = (V_2, E_2)$ are said to be *isomorphic* (written $G_1 \simeq G_2$) if there exists a one-to-one, onto mapping $f : V_1 \rightarrow V_2$, such that $uv \in E_1 \iff f(u)f(v) \in E_2$. That is, $G_1 \simeq G_2 \iff$ [Two vertices are adjacent in G_1 if and only if their images under f are adjacent in G_2]. Thus f preserves adjacency. The two graphs in Figure 1.6(a) are isomorphic because of the existence of the mapping $f(u_i) \equiv v_{i+1} \pmod 4$, for $i = 1, 2, 3, 4$.

It is interesting to attempt to characterize this concept of isomorphism. Isomorphic graphs have:

   (i) the same number of vertices,
   (ii) the same number of edges, and
   (iii) an equal number of vertices of any degree.

However, these properties are necessary but not sufficient criteria for isomorphism, as evidenced by the two graphs in Figure 1.6(b) which have properties (i), (ii), and (iii), but which are not isomorphic. An *automorphism* of a graph is an isomorphism of G with itself.

It is sometimes appropriate to examine just part of a graph. This can be done in a number of different ways. Given a graph $G = (V, E)$, we can consider: (i) just some of its vertices and all of the existing edges in E joining pairs of them, (ii) all of its vertices and just some of the existing edges in E joining pairs of them, or (iii) just some of its vertices and just some of the existing edges in E joining pairs of them, provided that we include all vertices incident with any edges considered.

More formally:
   (i) Let U be a nonempty subset of V. The graph whose vertex set is U and whose edge set comprises exactly the edges of E which join vertices in U is termed an *induced subgraph of G*.
   (ii) Let F be a (proper) subset of E. The graph (V, F), is termed a *(proper) spanning subgraph* of G.
   (iii) A structure formed by the combination of (i) and (ii) which results in a graph (i.e., the incident vertices of each selected edge are also selected) is termed a *subgraph* of G.

Figures 1.6(c), (d), and (e) provide examples of an induced subgraph, a spanning subgraph, and a subgraph, respectively, of the first graph shown in Figure 1.6(a). If G_1 is a subgraph of graph G_2, then G_2 is termed a *supergraph* of G_1.

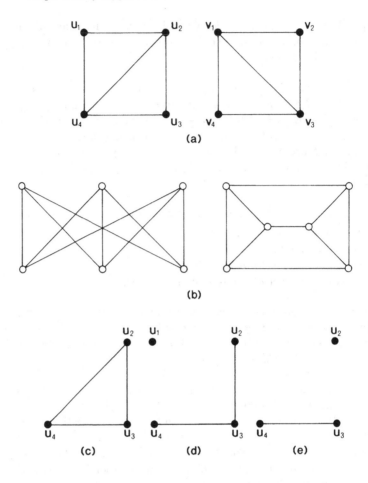

Figure 1.6 (a) Two isomorphic graphs (b) two non-isomorphic graphs
(c) an induced subgraph (d) a partial graph (e) a subgraph

When a vertex v (or the vertices of a subgraph G_1) and all the edges incident with it are removed from a graph G, the resulting graph is denoted by $G-v$ $(G-G_1)$. When an edge e, is removed from a graph, the resulting graph is denoted by $G-e$. Also if an edge e (set of edges E'), not already present in a graph G, is added to G then the resulting graph is denoted by $G+e$ $(G+E')$.

1.3 Summary

In this introductory chapter we have set the stage for the following material on graph theory and its applications. We have introduced a small number of both puzzles and practical problems which gave rise to the discovery of various graph theoretic concepts. These included: The Königsberg bridge problem, Hamilton's game, map colouring, and more serious applications in chemistry and physics.

We then went on to define rigorously some of the notions that arose during the considerations of these problems. Specifically, we defined the elementary concepts of graphs, multigraphs, pseudographs, digraphs, oriented and labelled graphs, isomorphisms, automorphisms, and subgraphs.

The purposes of this chapter are to give a flavour of graph theory as a branch of mathematics and to hint at its utility as a practical problem-solving tool in a variety of fields.

1.4 Exercises

1.1 Draw graphs representing the paraffins C_kH_{2k+2}, $k = 1, 2, 3, 4$. (See Section 15.2.)

1.2 Draw the 11 non-isomorphic graphs with four vertices.

1.3 Show that every graph with n vertices is isomorphic to a subgraph of the graph with n vertices, each pair of which is adjacent.

1.4 Use a graph theoretic approach to show that, in any set of n people ($n > 1$), there are at least two people with exactly the same number of friends within the set.

1.5 Prove that in any set of six people, there are either (i) three people who all know each other, or (ii) three people who are all strangers to each other.

1.6 Prove that the maximum number of edges in a graph with n vertices is $n(n - 1)/2$.

1.7 Add one bridge to Figure 1.1 making possible a tour of all bridges to be crossed once each, but in which the starting point is different from the finishing point. (See Section 4.2.)

1.8 Add a further bridge to the one added in Exercise 1.7 making possible a tour of all bridges to be crossed once each in which the starting point is coincident with the finishing point. (See Section 4.2.)

1.9 Find a tour of the edges of the graph of Figure 1.3 in which each vertex is traversed exactly once, which has the starting point coincident with the finishing point. (See Section 4.3.)

1.10 Draw a map with the minimum number of regions requiring four colours in order to colour it. (See Section 8.2.)

2 Connectivity

2.1 Introduction

One of the most important graph theoretic concepts is that of *connectivity*. We now introduce some of the ideas concerned with this aspect of graph structure. A *walk* in a graph G, is a sequence of vertices and edges of G of the form:

$$\langle v_1, \{v_1, v_2\}, v_2, \{v_2, v_3\}, v_3, \ldots, v_{n-1}, \{v_{n-1}, v_n\}, v_n \rangle.$$

Note that the sequence begins and ends with a vertex, and each edge is incident with the vertices immediately preceding and succeeding it. The walk is said to *join* v_1 and v_n. When there is no ambiguity, the walk is denoted by:

$$\langle v_1, v_2, v_3, \ldots, v_n \rangle.$$

A walk is termed *closed* if $v_1 = v_n$, and *open* otherwise. A walk is termed a *trail* if all of its edges are distinct, and a *path* if all of its vertices (and necessarily all of its edges) are distinct. A closed walk with at least three

17

vertices, and all of its vertices distinct, (except for the first and the last) is called a *cycle*. A cycle is termed *even* if it contains an even number of edges (and necessarily vertices), and *odd* otherwise. A cycle with n edges is termed an *n-cycle*. If $n = 3$, it is termed a *triangle*. A graph that is solely an n-cycle is denoted by C_n. A graph G, is termed *connected* if every pair of vertices in G are joined by a path and is termed *disconnected* if it is not connected. A graph is termed *k-connected* if at least k vertices must be removed to render G disconnected. A graph without cycles is termed *acyclic*. A maximal connected subgraph of a graph G is called a *component* of G. A graph in which every pair of its n vertices are directly connected by an edge is termed *complete*, and is denoted by K_n.

A vertex v, (edge e) of a graph G is termed a *cut-vertex* (*bridge*) of G if $G–v$ ($G–e$) comprises a greater number of components than does G. Thus if G is connected (and thus comprises exactly one component) with a cut-vertex v, (bridge e), then $G–v$, ($G–e$) must, by definition, comprise at least two components. As examples, the graphs in Figure 2.1 (a) and (b) have vertex v as a cut-vertex.

A *nonseparable graph* is a connected graph with no cut-vertices. As examples, the graphs in Figure 2.2 are nonseparable. All graphs which are not nonseparable are termed *separable*. This means that any disconnected graph is separable.

A *block* of a separable graph G is a maximal nonseparable subgraph of G. As examples, the subgraphs induced by the vertex set $\{u, v, w, x\}$, are blocks of the graphs in Figures 2.1 (a) and (b).

Consider now the two non-isomorphic graphs G_1 and G_2, shown in Figures 2.1 (a) and (b). They both have a single cut-vertex, namely v. If this cut-vertex is split into two vertices in each of G_1 and G_2 we obtain the edge-disjoint graphs G_1^1 and G_2^1, as shown in Figures 2.1 (c) and (d). It is evident that G_1^1 and G_2^1 are isomorphic. Graphs which become isomorphic after the splitting of all of their cut-vertices are said to be *1-isomorphic*.

There is a further generalization of isomorphism which we shall find useful in the analysis of duality, carried out in Chapter 5. It is illustrated via the non-isomorphic graphs in Figure 2.2. Two graphs G_1 and G_2, are said to be *2-isomorphic* if there exists a one-to-one correspondence between their edge sets such that the edges of a cycle in G_1 correspond to the edges of a cycle in G_2 and vice versa.

The number of edges incident with a vertex v, in a graph is called the *degree* (or *valence*) of v, and is written $d(v)$. The number of arcs directed away from a vertex v, in a digraph is called the *outdegree* of v and is written $od(v)$. The number of arcs directed towards a vertex v, in a digraph is

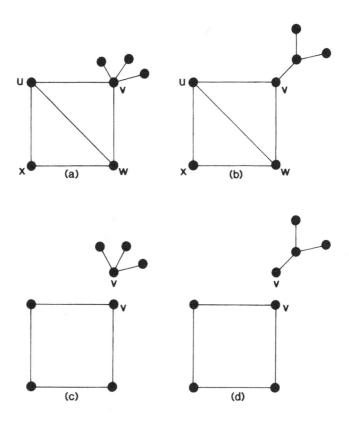

Figure 2.1 1-isomorphic graphs

called the *indegree* of v and is written $id(v)$. For any vertex v in a digraph we define $d(v) = od(v) + id(v)$.

2.2 Elementary Results

Theorem 2.1

In any graph there is an even number of vertices of odd degree.

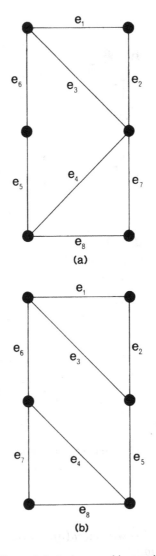

Figure 2.2 2- isomorphic graphs

Proof

The sum of the degrees of all vertices in any graph is made up of two sums:
the degree sum of vertices of even degree and the degree sum of vertices of

odd degree. Hence the second sum can be seen as the difference between two sums which are even and hence it must itself be even. The result follows from the fact that an even sum comprising solely odd terms must have an even number of terms. ∎

A graph in which all vertices are of equal degree is termed *regular*. A vertex of degree 1 is termed *pendant*. An edge incident with a pendant vertex is itself termed *pendant*. A vertex of degree 0 is termed *isolated*.

A graph (V, E), in which $E = \emptyset$ is termed *null*. (Recall that V cannot be empty in any graph). A vertex of degree greater than 1 is termed *internal*.

Theorem 2.2

If a graph has exactly two vertices of odd degree they must be connected by a path.

Proof

Let G be a graph with all its vertices of even degree, except for v_1 and v_2 which are of odd degree. Consider the component C, to which v_1 belongs. By Corollary 2.1, C has an even number of vertices of odd degree. Thus C must contain v_2, the only other vertex of odd degree. Hence v_1 and v_2 are in the same component. As a component is, by definition, connected there must exist a path between v_1 and v_2. ∎

The following lemma is needed in the proof for Theorem 2.3, given later.

Lemma 2.1

For any set of positive integers: $n_1, n_2, \ldots, n_k$:

$$\sum_{i=1}^{k} n_i^2 \leq \left(\sum_{i=1}^{k} n_i \right)^2 - (k-1)\left(2\sum_{i=1}^{k} n_i - k \right)$$

Proof

$$\sum_{i=1}^{k} (n_i - 1) = \sum_{i=1}^{k} n_i - k$$

Therefore

$$(n_1 - 1)^2 + (n_2 - 1)^2 + \ldots + (n_k - 1)^2 + \sum_{i=1}^{k} \sum_{\substack{j=1 \\ i \neq j}}^{k} (n_i - 1)(n_j - 1)$$

$$\ldots = \left(\sum_{i=1}^{k} n_i \right)^2 - 2k \sum_{i=1}^{k} n_i + k^2$$

Therefore

$$\sum_{i=1}^{k} n_i^2 - 2\sum_{i=1}^{k} n_i + k \le \left(\sum_{i=1}^{k} n_i\right)^2 - 2k\sum_{i=1}^{k} n_i + k^2$$

Therefore

$$\sum_{i=1}^{k} n_i^2 \le 2\sum_{i=1}^{k} n_i - k + \left(\sum_{i=1}^{k} n_i\right)^2 - 2k\sum_{i=1}^{k} n_i + k^2.$$

Thus the result follows. ∎

Theorem 2.3

A graph with n vertices and k components cannot have more than $\frac{1}{2}(n - k)(n - k + 1)$ edges.

Proof

Let n_i = the number of vertices in component i, $1 \le i \le k$.

Then

$$\sum_{i=1}^{k} n_i = n.$$

A component with n_i vertices will have the maximum possible number of edges when it is complete. That is, it will contain $\frac{1}{2}n_i(n_i - 1)$ edges.

Hence the maximum number of edges is:

$$
\begin{aligned}
\frac{1}{2}\sum_{i=1}^{k} n_i(n_i - 1) &= \frac{1}{2}\sum_{i=1}^{k} n_i^2 - \frac{1}{2}\sum_{i=1}^{k} n_i \\
&\le \frac{1}{2}[n^2 - (k-1)(2n-k)] - \frac{1}{2}n \qquad \text{by Lemma 2.1} \\
&= \frac{1}{2}[n^2 - 2nk + k^2 + n - k] \\
&= \frac{1}{2}(n - k)(n - k + 1).
\end{aligned}
$$

∎

2.3 Structure Based on Connectivity

The *complement* $\overline{G}$, of a graph $G = (V, E)$, is the graph with the vertex set V such that two vertices are adjacent in $\overline{G}$ if and only if these vertices are not adjacent in G. A graph G is said to be *self-complementary* if $G \simeq \overline{G}$.

Definition

A graph $G = (V, E)$ is *n-partite*, $n > 1$, if it is possible to partition V into n subsets: $V_1, V_2, \ldots, V_n$ (called *partite sets*) such that every edge of E joins a vertex of V_i to a vertex of $V_j, i \neq j$.

When $n = 2$, n-partite graphs are called *bipartite graphs*. These graphs are of particular utility and will be discussed later in this book. Two isomorphic bipartite graphs on six vertices are shown in Figure 2.3.

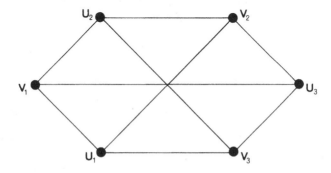

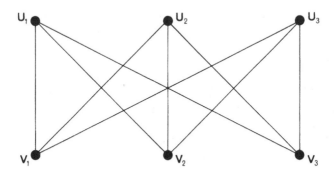

Figure 2.3 Two isomorphic bipartite graphs

Theorem 2.4

A graph is bipartite if and only if it contains no odd cycles.

Proof

Let $G = (V, E)$ be a connected bipartite graph with bipartite sets V_1 and V_2.

Suppose there exists a cycle: $\langle v_1, v_2, \ldots, v_k, v_1 \rangle$ in G.

(Necessity)

Suppose $v_1 \in V_1$. Then v_i is a member of V_1 if i is odd, and v_i belongs to V_2 if i is even. This implies that k is even. Hence the result.

(Sufficiency)

Consider a vertex $v_1 \in V$. Define V_1 to be the subset of V comprising v_1 and all vertices of G whose shortest paths from v_1 (in terms of the number of edges) contain an even number of edges. Let V_2 comprise the remaining vertices of G. Because every cycle in G is even, every edge of G joins a vertex in V_1 to a vertex in V_2. This is so because, suppose there exist two adjacent vertices, say v_2 and v_3, in V_1. Then the cycle C, say, constructed by adjoining the shortest path from v_1 to v_2, the edge $v_2 v_3$, and the shortest path from v_3 to v_1, is odd. (Here the shortest paths are defined in the sense introduced earlier.) This is a contradiction. Hence the result follows. ∎

We return now to the concept of the complement of a graph which was introduced earlier. We state a number of results concerning complementarity.

Theorem 2.5

Any graph and its complement cannot both be disconnected.

Proof

Suppose that u and v are two vertices of $\overline{G}$, the complement of a disconnected graph G. If u and v belong to different components of G, then u and v are adjacent in $\overline{G}$ by definition. If u and v belong to the same component say G_i, of G, then let t be a vertex of some other component, say G_j, of G. By definition, both u and t, and v and t are adjacent in $\overline{G}$. In either case, u is connected to v by a path in $\overline{G}$. Thus $\overline{G}$ is connected. ∎

2.4 Summary

In this chapter we have discussed a basic graph theoretic concept, namely *connectivity*. It is fundamental in the study of graphs as mathematical objects. It is also vital in many applications involving graphs. We have introduced the notions of walks, trails, paths, cycles, connectivity, non-separability, blocks, isomorphism, vertex degree, partiteness, and complements. We then went on to introduce some elementary results concerning these ideas. We concluded with a brief discussion on structure based on connectivity. An algorithm to detect whether or not a graph is connected is stated in Section 9.5.2.

2.5 Exercises

2.1 Let p and q be any two vertices of a connected graph G. Prove that there must be a walk between p and q containing all the vertices of G.

2.2 Prove that "is connected to" is an equivalence relation on the set V, of any graph $G = (V, E)$.

2.3 Characterize those graphs G, having the property that every induced subgraph of G is a connected subgraph of G.

2.4 Prove that any two connected graphs having the same number of vertices, all of which are of degree two, are isomorphic.

2.5 The *vertex (edge) connectivity* of a connected graph G, is defined as the minimum number of vertices (edges) whose removal from G leaves the remaining graph disconnected. Show that a connected graph with n vertices and vertex connectivity k must have at least $kn/2$ edges.

2.6 Prove that every complete graph with n (> 1) vertices has at least $(n - 1)([n/2] - 1)$ cycles.

2.7 Prove that a graph with n vertices must be connected if it has more than $(n - 1)(n - 2)/2$ edges.

2.8 Prove that a graph with n vertices $(n > 2)$ cannot be bipartite if it has more than $n^2/4$ edges.

2.9 Prove that every graph with n vertices is isomorphic to a subgraph of K_n.

2.10 Prove that if a graph has more edges than vertices then it must possess at least one cycle.

2.11 Prove that if a graph has at least two more vertices than edges then it is disconnected.

2.12 Suppose G is a graph with n vertices and m edges and that each vertex in G has degree either p or $p + 1$. Let the number of vertices of degree p be n_p and the number of degree $p + 1$ be $n_p + 1$. Prove that $n_p = (p + 1)n - 2m$.

3 Trees

"Woodman, spare that tree!"

G.P. Morris

Trees are among the most important graphs. This is mainly due to the fact that many of the applications of graph theory, directly or indirectly, involve trees. A good illustration of this is the application in molecular evolution, which is covered in Section 15.3. We begin this chapter by introducing the concept of a tree and its properties. Many applications of trees will become self-evident once the reader has grasped a few simple notions. Later in the present chapter we go on to introduce various results concerning trees, distances between trees, binary trees, tree enumeration, spanning trees, and fundamental cycles.

3.1 Characterizations

Definition

A *tree* is a connected acyclic graph. Its edges are called *branches*.

To give the reader an instant feel for what type of graphs are trees, we list in Figure 3.1, all the trees with at most seven vertices. Before studying this figure in detail, the reader is urged to attempt to draw all the trees with exactly eight vertices.

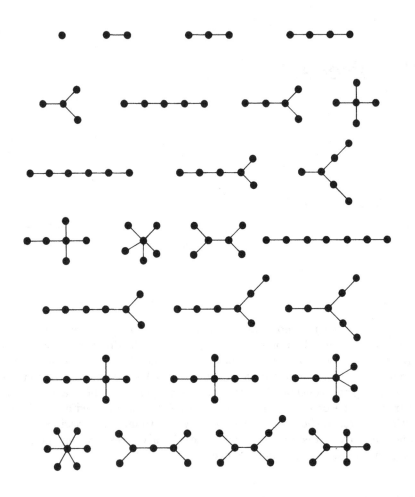

Figure 3.1 All the trees with at most 7 vertices

Having studied the definition and Figure 3.1, many instances in society and nature where trees can be used to illustrate common activities may come to mind to the reader. It is natural and useful to use trees to model kinship relations from sociology, sequential decision or sorting activities, minimally connected communications networks, and the anatomy of various structures with a number of levels, as well as numerous other phenomena from a wide variety of fields. A number of elementary observations can be made about trees:

(i) The removal of any single edge from a tree creates a graph which is disconnected.

(ii) In any connected graph which is not a tree, it is possible to find at least one edge, whose removal will leave the graph still connected.

As a result of the above two observations, a tree can be characterized as a minimally connected graph in the sense that it does not possess a proper spanning subgraph which is connected.

(iii) There is exactly one path between every pair of vertices in a graph G if and only if G is a tree.

The necessity of observation (iii) follows because the existence of a path between every pair of vertices implies that G is connected. If G contained a cycle then there would exist at least one pair of vertices in G which had the property that there was at least two distinct paths between them. As this is not the case, G is acyclic. Therefore G is a tree.

The sufficiency of observation (iii) follows because, as G is a tree, it is (by definition) connected. Therefore there is at least one path between every pair of its vertices. If there exists a pair of vertices in G with the property that there is at least two distinct paths between them then the union of these two paths will produce a cycle. This cannot be the case as G is a tree and hence acyclic. Therefore there is exactly one path between every pair of vertices in G.

(iv) A connected graph G is a tree if and only if G has one less edge than it has vertices.

Observation (iv) can be proved very easily by mathematical induction.

(v) An acyclic graph which has one less edge than it has vertices must be connected.

The proof of observation (v) follows easily by contradiction.

By using the above observations it is possible to specify a number of alternative definitions for a tree:

T(i)   A tree is a connected graph with n vertices and $(n-1)$ edges.

T(ii)   A tree is an acyclic graph with n vertices and $(n-1)$ edges.

T(iii) A tree is a graph in which there is exactly one path between every pair of its vertices.

T(iv)   A tree is an acyclic graph which has the property that if any two of its vertices which are not adjacent are joined directly by an edge then the resulting graph possesses exactly one cycle.

3.2 Theorems on Trees

Theorem 3.1

Any tree with at least two vertices has at least two pendant vertices.

Proof

Let the number of vertices in a given tree be n (> 1). From Section 3.1 we have that the tree has $(n - 1)$ edges. Thus the degree sum of the tree is $2(n - 1)$. This degree sum is to be divided among the n vertices. Because a tree is connected it cannot possess a vertex of 0 degree. Therefore each vertex contributes at least 1 to the above sum. Therefore there must be at least two vertices of degree exactly 1. ■

Definition

An acyclic graph is termed a *forest*.

The components of a forest are trees. Removal of a single edge from a tree creates a forest of exactly two trees. It is suggested that the reader establishes why exactly two trees are created. Removal of a further edge will create a forest of three trees. Continuing in this way the removal of any $(k - 1)$ edges from a tree will create a forest of k trees. The ultimate product of this removal process is a forest of n trees, each being an isolated vertex. Thus we have:

Theorem 3.2

A forest of k trees which has a total of n vertices has $(n - k)$ edges.

Proof

The proof follows from the observation immediately preceding the statement of the theorem. ■

3.3 Tree Distances

In studying the very last graph in Figure 3.1, it can be seen that the two vertices with degree greater than 1 are somewhat more central, in a certain sense, than the pendant vertices. Before putting this concept of centrality on a firm mathematical footing, we must develop the concept of distance in a graph.

Definition

The *distance* d_{uv}, between two vertices u and v of a graph G, is equal to the number of edges in the path with the shortest number of edges connecting u and v.

As an example, the distance between a pair of distinct vertices selected at random from the graph of Figure 3.1 is 1, or 2, or 3. As a consequence of our previous observations, there is exactly one path between any two vertices in a tree. This makes these distances relatively easy to calculate for trees. Because the distance function we have defined, $d : V \times V \rightarrow \Re \cup \{0\}$, is non-negative, symmetric, and obeys the triangle inequality, it is a metric in the normal topological sense.

We are now in a position to develop rigorously the concept of centrality.

Definitions

The *eccentricity* $e(v)$, of a vertex v of a connected graph $G = (V, E)$, is defined as $\max_{u \in V} d_{uv}$.

The *radius* $r(G)$, of a connected graph G is equal to the eccentricity of the vertex in G which is of minimum eccentricity.

The *diameter* $d(G)$, of a connected graph G is equal to the eccentricity of a vertex in G which is of maximum eccentricity.

A vertex v, in a connected graph G is termed *central* if $e(v) = r(G)$.

The *centre* of a connected graph G is the set of all central vertices in G.

Each vertex is labelled with its eccentricity in the top tree in Figure 3.2. As can be seen, this graph has a centre comprising two vertices, each labelled C. Clearly a graph can have a centre with cardinality greater than 1. Indeed a graph can have an arbitrary number of central points. For example, every vertex is a central point in C_n. The fact that the tree in the Figure 3.2 has two central points begs the question as to how many central points there can possibly be in any tree. The following theorem provides the answer.

Theorem 3.3

Every tree has a centre comprising either one vertex or two adjacent vertices.

Proof

The result is clear for the first two trees in Figure 3.1. Our strategy is to show that any tree T has the same centre as a tree T', obtained from it by removing from T all of its pendant vertices. Naturally the eccentricity of any vertex v, in T is equal to the distance from v to some pendant vertex in T. Therefore the eccentricity of each vertex in T' is one less than the eccentricity of that vertex in T. Therefore the vertices of minimum eccentricity in T will have minimum eccentricity in T'. Therefore T and T' have the same centre. Let us successively remove the pendant points all at once from T. If this process is repeated successively, a sequence of trees will result all of which have the same centre, and indeed have the same centre as T. This process leads to one of two results: either the first or the second of the two trees in Figure 3.1 eventually occur. In either case the vertices of this final tree comprise the centre of T. Hence the result. ■

This process is illustrated in the lower diagrams of Figure 3.2. By definition, the radius of the original tree in Figure 3.2 is 4. The reader should confirm that the diameter of that tree is 7. Thus we see that the diameter of a tree is not necessarily twice its radius. Further properties of the diameter of a tree are given in Exercises 3.13 and 3.14.

3.4 Binary Trees

We turn now to a brief discussion of concepts which are essential to the application of trees in molecular evolution, as discussed in Section 15.3.

Definitions

A tree in which exactly one vertex is distinguished is termed a *rooted tree*.

The distinguished vertex in a rooted tree is termed its *root*.

All the rooted trees with five vertices are shown in Figure 3.3, with the root distinguished by a square rather than a disk. Sometimes trees are turned into digraphs by assigning an orientation to each edge away from the root, to make it an arc. Ordinary trees are occasionally called *free* to underline the fact that they do not possess a root. One of the most common classes of rooted trees is termed *binary*.

Definition

A *binary tree* is a rooted tree with at least three vertices in which the root is of degree 2 and all other internal vertices are of degree 3.

We now state some theorems about binary trees.

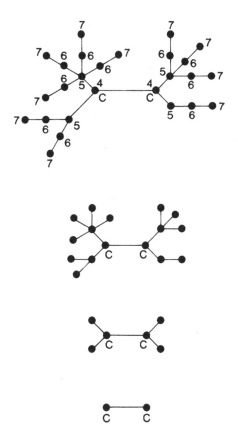

Figure 3.2   Trees with two central points

Theorem 3.4

Every binary tree has an odd number of vertices.

Proof

Apart from the root, every vertex in a binary tree is of odd degree. By Corollary 2.1, there must be an even number of such odd vertices. Therefore when the root (which is of even degree) is added to this number the total number of vertices must be odd. ∎

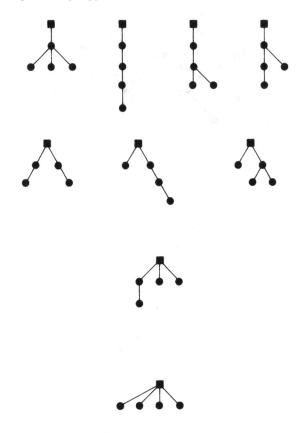

Figure 3.3 The rooted trees with 5 vertices

Theorem 3.5

There are $\frac{1}{2}(n+1)$ pendant vertices in any binary tree with n vertices.

Proof

Suppose that T is a binary tree with n vertices. Let m be the number of pendant vertices in T. There are then $(n-m)$ internal vertices in T and $(n-m-1)$ vertices of degree 3. Thus there are $\frac{1}{2}(3(n-m-1)+2+m)$ edges in T. However since T has n vertices, it has $(n-1)$ edges. In equating these two expressions and solving for m the result follows. ∎

3.5 Tree Enumeration

3.5.1. Introduction

As we noted in Section 1.1.4, Cayley (1857, 1874, 1889, 1895, 1896) introduced the concept of a tree, and enumerated all rooted trees in 1889. Since that time, enumerative methods for counting various classes of graphs, including trees, have been developed, but are still far from completely scientific. The father of this field was Redfield (1927) whose contributions foreshadowed much of the later work by Polya (1937, 1940). It is not the purpose of this chapter to provide a survey of counting methods. Instead we state a few results and illustrate the power of current enumeration methods by referring to the counting of one specialized class of trees. This class will be discussed when the application of graph theory in molecular evolution is introduced in Section 15.3. A further illustration of graph enumeration, involving the counting of maximal planar graphs, defined in Section 5.3, is given in Section 11.4.1.

Theorem 3.6

There are $2^{\binom{n}{2}}$ labelled graphs with n vertices.

Proof

Consider a graph with its vertices labelled $1, 2, \ldots, n$. In any such graph, each of the $\binom{n}{2}$ possible edges is either present or absent. Hence the result. ∎

Cayley (1889) stated the corresponding result for the special class comprising labelled trees:

Theorem 3.7

There are n^{n-2} labelled trees with n vertices.

Proof

The proof is too involved to be given here, but can be found in Moon (1967).

3.5.2* Enumerating Phylogenetic Trees

Biologists often represent postulated evolutionary relationships between existing biological species by means of a tree. Such a diagram, linking related species to a common ancestor, is called a *phylogenetic tree* or *phylogeny*.

Definition

A *phylogeny* is a tree $T = (V, E)$, together with a set $\{1, 2, \ldots, n\}$ of labels, and a function f, mapping the labels into V, where every vertex of degree less than three must be in the image of f.

This definition allows for the possibilities that some vertices may possess multiple labels and some vertices no labels at all. The labelling set corresponds to a given set of existing species.

Definitions

The *magnitude* of a phylogeny is defined to be n, the number of its labels.

The *order* of a phylogeny is defined to be its number of vertices.

A *planted phylogeny* is a rooted tree which is a phylogeny having a pendant vertex which is distinguished, and is termed its *root*.

This root represents the common ancestor of all the species in the labelling set and thus is not given one of the labels of the set, and is not counted in calculating the order of the phylogeny. In the search for phylogenies which are optimal, in the sense of satisfying criteria of scientific models of evolution, it is of interest to know how many possible feasible phylogenies exist for a given set of n labelled species. However there are various classes of phylogenies. They can be required to:

  (i) be binary,
  (ii) have all vertices of degree three or more unlabelled,
  (iii) have only singleton labels,
  (iv) have no vertices of degree two.

Combinations of the requirement or absence of each of the above for conditions, ostensibly creates 16 classes of phylogenies. However four classes are empty as there are no vertices of degree two (other than the root) in a binary phylogeny. Hence there are 12 nonempty classes, as defined in Table 3.1.

The numbers of phylogenies for all 12 classes can be calculated exactly for given n, along with the mean and variance of the number of vertices in the phylogenies, and the asymptotic behaviour of the exact numbers, by using the methods of Harary et al. (1975). This has been done by Foulds and Robinson (1980, 1981, 1984, 1985, and 1988) .

Class	Must be Binary	Only Vertices of Degree 1 or 2 Labelled	All Labels Singleton	Vertices of Degree 2 Allowed
1	NO	NO	NO	YES
2	NO	NO	YES	YES
3	YES	YES	YES	NO
4	YES	YES	NO	NO
5	YES	NO	YES	NO
6	YES	NO	NO	NO
7	NO	YES	YES	NO
8	NO	YES	NO	NO
9	NO	NO	YES	NO
10	NO	YES	YES	YES
11	NO	NO	NO	NO
12	NO	YES	NO	YES

Table 3.1. The 12 classes of phylogenies.

3.6 Spanning Trees

We shall now consider trees as subgraphs of larger graphs. Consider a connected graph $G = (V, E)$, which contains a subgraph which is the tree $T = (V', E')$. The edges E', of T are called *branches* and the edges of G which are not in T are called *chords* (both relative to T).

Definition

If $V' = V$ then T is said to be a *spanning tree* of the graph G.

That is, a spanning tree contains all the vertices of the graph of which it is a subgraph. It is also clear that only connected graphs have spanning trees as subgraphs. Of course, if a connected graph G, has a unique spanning tree then G is itself a tree and that spanning tree is G itself. Further, each component of an arbitrary graph G will contain at least one spanning tree. A collection of such spanning trees, one for each component of G, is termed a *spanning forest*. This is illustrated in Figure 3.4. Each spanning tree in the spanning forest is depicted by bold lines. The following theorem provides the number of chords in any graph with a spanning tree:

Theorem 3.8

A connected graph containing n vertices and e edges will have, relative to any of its spanning trees, $(e - n + 1)$ edges that are not members of that spanning tree.

Proof

The proof follows immediately from the definition of a spanning tree and from the fact that a tree with n vertices has $(n - 1)$ edges. ∎

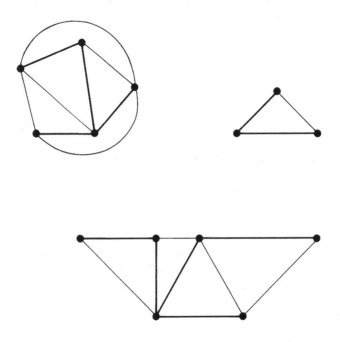

Figure 3.4 A spanning forest

We now consider the important concept of the fundamental cycle, which is one of the building blocks by which graphs can be created and transformed.

3.7 Fundamental Cycles

In the last section we discussed the identification of a spanning tree in a connected graph by the successive removal of edges from its cycles. There is the obvious corollary that if we begin with a connected acyclic graph, namely a tree, we can add an edge to join two of its vertices that are not directly connected, to create a cycle. This fact coincides with definition T(iv) of a tree, given in Section 3.1. Let us now assume that the tree T, is a spanning subgraph of a connected graph G, with n vertices and e edges. We can identify the chords of G with regard to T. The addition of a chord to T will create exactly one cycle with regard to T, called a *fundamental cycle*.

Thus, with regard to any of its spanning trees, the graph G has as many fundamental cycles as it has chords, namely $(e - n + 1)$. Each spanning tree of G will have associated with it a set of $(e - n + 1)$ fundamental cycles. We shall now illustrate these ideas via Figure 3.5. At the top of the figure is a connected graph and directly below it are two of its spanning trees with edge set: $\{d, a, b\}$ and $\{d, c, b\}$ respectively. Adding the chord e, to the first spanning tree creates the cycle $\langle a, b, e \rangle$. Adding the other chord c, to that spanning tree creates the fundamental cycle $\langle a, b, c, d \rangle$. Repeating an analogous process for the second spanning tree, we create a different set of fundamental cycles. Namely the cycles: $\langle a, b, c, d \rangle$ and $\langle e, c, d \rangle$. These sets of fundamental cycles have exactly one cycle in common namely $\langle a, b, c, d \rangle$. Thus we see that the set of fundamental cycles is very much dependent on the spanning tree selected. As an exercise, the reader should identify the other six spanning trees for the original graph and the set of fundamental cycles that each one creates. Because the original graph had two chords, there will always be two fundamental cycles in the set of fundamental cycles created by any spanning tree. The number of fundamental cycles is constant for any connected graph.

There are many applications in which we want to analyse the cycles of a connected graph. The most straightforward way to do that is to create a spanning tree and to create a set of fundamental cycles from it. It turns out that all the cycles of a graph can be created by performing a special type of simple matrix arithmetic on the elements in the set of fundamental cycles. This fact was discovered by Kirchhoff, in his analysis of electrical networks and was discussed briefly in Section 1.1.5. These ideas will be explored in Section 6.3 on fundamental cycle matrices, and in Section 13.1 on electrical circuit analysis.

3.8 Summary

In this chapter we have introduced the class of graphs called trees, which are of great interest in their own right and also for their applications. One of the reasons for their importance from a mathematical point of view is that often a tree represents a simple type of graph which is worth analysing first

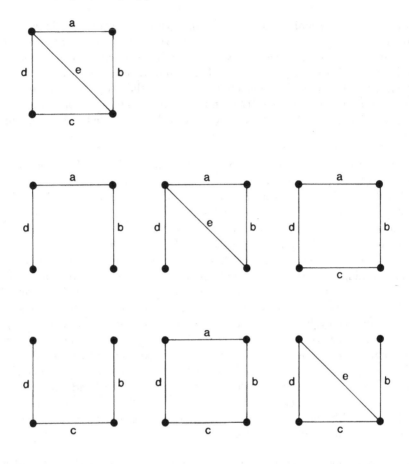

Figure 3.5 Fundamental cycles in a connected graph

in order to judge whether or not a particular proposition is likely to be true. Trees have numerous applications in a wide variety of practical fields. In order to provide the tools for some of these applications we have introduced some elementary theorems concerning trees, tree distances, binary trees, tree enumeration, spanning trees, and fundamental cycles. These concepts will be used throughout the rest of this book.

3.9 Exercises

3.1 Draw the 23 unlabelled trees with 8 vertices.

3.2 Draw the 20 rooted trees with 6 vertices.

3.3 Draw a tree which has radius 5 and diameter 10.

3.4 Prove the observations, and justify the validity of, the alternative definitions stated in Section 3.1.

3.5 Prove that any subgraph G', of a connected graph G, is contained in a spanning tree of G if and only if G' is acyclic.

3.6 Prove that every connected graph possesses a spanning tree as a subgraph.

3.7 A *star* is a tree which has at most one non-pendant vertex. Show that a tree with more than two vertices is a star if and only if it has diameter two.

3.8 How many spanning trees does K_4 have?

3.9 Prove that each spanning tree of a connected graph G, contains all the pendant edges of G.

3.10 Prove that each edge of a connected graph G, belongs to at least one spanning tree of G.

3.11 A connected graph $G = (V, E)$, is said to be *minimally connected* if, for each edge $e \in E$, $G-e$ is disconnected. Prove that a connected graph is minimally connected if and only if it is a tree.

3.12 Prove that a subgraph of a connected graph G, is a maximal acyclic subgraph of G if and only if it is a spanning tree of G.

3.13*Under what conditions is the diameter of a tree equal to twice its radius?

3.14 Prove that the diameter of any tree is equal to the number of edges in its path with the maximum number of edges.

4 Traversability

"To travel hopefully is a better thing than to arrive."

El Dorado, R.L. Stevenson

4.1 Introduction

In Chapter 1 we met two puzzles that could be modeled by graphs: the Königsberg bridge problem and Hamilton's game. There are many other games and puzzles which can be profitably analysed by using graph theoretic concepts, as will be seen in Section 11.5. This is because many puzzles and games can be converted into an equivalent graph theoretic problem in which the solution can be found by attempting to construct either an Eulerian trail or a Hamiltonian cycle within an appropriate graph. We begin with Eulerian graphs.

4.2 Eulerian Graphs

A general question arises naturally from the discussion of the Königsberg bridge problem in Section 1.1.1. That is, given a connected graph G, is it possible to find a closed trail containing every edge of G exactly once? Such a trail is termed an *Euler trail*. (An open trail containing every edge of G exactly once is termed on *open Euler trail*.) Many applications involve finding Eulerian trails in multigraphs rather than in graphs. Thus in this section we provide definitions and results for multigraphs which are analogous to those for graphs.

Definitions

A multigraph containing an Euler trail is said to be *Eulerian*.

A multigraph containing an open Euler trail is termed *unicursal*.

We now provide two alternative characterizations of Eulerian multigraphs.

Theorem 4.1

The following statements are equivalent for any connected multigraph G.

G is Eulerian.

Every vertex of G is of even degree.

The set of edges of G can be partitioned into cycles.

Proof

We first show that if a connected multigraph G is Eulerian then every vertex of G has even degree. Let G be an Eulerian multigraph. Let T be an Euler trail in G. In tracing this trail, each time a vertex is met it contributes 2 to the degree of that vertex. Because each edge of G occurs exactly once in T, every vertex must have even degree.

We now show that if every vertex of a connected multigraph G has even degree then the set of edges of G can be partitioned into cycles. Suppose that G is a connected multigraph in which every vertex has even degree. As G is connected and contains at least one vertex, every vertex has degree at least 2. Therefore G contains a cycle, say C. The removal of the edges of C results in a spanning subgraph G', say, in which every vertex has even degree. If G' has no edges then the result follows. Otherwise if G' still has edges, the above process is repeated on G'. This results in a graph in which all of the vertices are still even. This process is repeated until a multigraph with no edges is obtained. Each application of the above process creates a cycle and the totality of these cycles represents a partition of the edges of G into disjoint cycles.

We finally show that if the set of edges in a connected multigraph G can be partitioned into cycles, then G is Eulerian. Suppose that G is a connected multigraph such that its edges can be partitioned into cycles. Let C_1 be one of the cycles in this partition. If G comprises solely C_1, then G is Eulerian. If not, there exists another cycle C_2, say, with a vertex v, in common with C_1. At least one such cycle C_2 must exist because G is connected. The walk which begins at v, and comprises the cycles C_1 and C_2 in succession, is a closed trail comprising the edges of these two cycles. By continuing this argument we construct a closed trail containing all the edges of G. Thus the result follows. ■

One of the conditions of Theorem 4.1 provides a very useful characterization of Eulerian multigraphs. Namely we can easily check to see whether a multigraph is Eulerian or not by testing to see whether or not all of its vertices are of even degree. If this is so, the multigraph is Eulerian. If not, then the multigraph is not Eulerian. Returning to the Königsberg bridge problem of Section 1.1.1, it then follows that, because this multigraph has vertices of odd degree, it is not Eulerian. Thus it is impossible to find the required trail. Present day Königsberg is now part of the Soviet Union and is called Kaliningrad. Two additional bridges have been built since Euler's day. Referring to Figure 1.1, the first was built between regions I and III. This leads to vertices I and III being adjacent in Figure 1.2, as shown in Figure 4.1 (a). It is still impossible to find an Euler trail in the new multigraph because vertices II and IV are still of odd degree. However we can ask a weaker question. Is it possible to find an open Euler trail containing all of the edges of the graph exactly once? The following theorem provides the answer in general.

Theorem 4.2

Let G be a connected multigraph. Then G contains an open Euler trail if and only if G has exactly two vertices of odd degree. If such a trail exists, it begins at one of these vertices of odd degree and terminates at the other vertex of odd degree.

Proof

This follows as an immediate corollary of the proof of Theorem 4.1. ■

This result can be used to show that the multigraph of Figure 4.1(a), is unicursal. Any open Euler trail in the multigraph must connect vertex II to vertex IV. We mentioned that two additional bridges have been built. The second directly connects Kneiphof island (II) to region (IV), in Figure 1.1. This adds a second edge to the vertices II and IV in Figure 1.2. The new multigraph is shown in Figure 4.1(b). Because all of its vertices are of even degree it is now possible to construct an Euler trail, such as that given by the sequence of edges: $\langle 1, 5, 8, 2, 6, 7, 9, 4, 3 \rangle$. Thus the Kaliningrad bridge problem now has a solution.

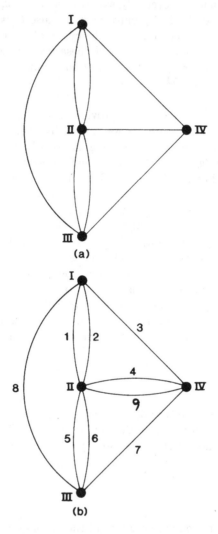

Figure 4.1 The graphs of the Kaliningrad bridge problem

4.3 Hamiltonian Graphs

In Section 1.1.2 we briefly discussed Hamilton's game in which one is asked to find a cycle which contains each vertex of a given graph. In order to study the general problem of finding such cycles in arbitrary graphs we must first put these ideas on a more rigorous foundation. To this end we begin with a few definitions.

Definitions

A graph with a spanning cycle is termed a *Hamiltonian graph*.

A spanning cycle in a graph is termed a *Hamiltonian cycle*.

Hamilton posed the question of establishing whether or not a given graph is Hamiltonian. Many mathematicians over the decades have unsuccessfully attempted to find an elegant characterization of Hamiltonicity. As yet there is no simple characterization of Hamiltonian graphs equivalent to Theorem 4.1 for Eulerian multigraphs. We shall now discuss sufficient conditions for a graph to be Hamiltonian. If a graph G is Hamiltonian then so is the graph $G + uv$, where u and v are distinct nonadjacent vertices of G. This trivial observation leads us to our first theorem concerning Hamiltonian graphs.

Theorem 4.3

Consider a connected graph $G = (V, E)$, with n vertices, where $n > 2$. Let u and v be a pair of distinct nonadjacent vertices of G such that:

$$d(u) + d(v) \geq n.$$

Then $G + uv$ is Hamiltonian if and only if G is Hamiltonian.

Proof

The necessity of the above proposition is obvious. We turn now to the sufficiency. Suppose that G is a graph with n vertices containing nonadjacent vertices u and v, such that $G + uv$ is Hamiltonian. Suppose that in G:

$$d(u) + d(v) \geq n.$$

Assume that G is not Hamiltonian. By assumption, there exists a Hamiltonian cycle of $(G + uv)$ containing the edge uv. Thus there is a path P, in G from:

$$u \text{ to } v : \langle u = u_1, u_2, \ldots, u_n = v \rangle \text{ in } G,$$

containing every vertex of G. If $u_1 u_i \in E$, $2 \leq i \leq n$, then $u_{i-1} u_n \notin E$, for otherwise

$$\langle u_1, u_i, u_{i+1}, \ldots, u_n, u_{i-1}, u_{i-2}, \ldots, u_1 \rangle$$

is a Hamiltonian cycle of G. Hence for each vertex of $\{u_2, u_3, \ldots, u_n\}$ adjacent to u_1 there is a vertex of $\{u_1, u_2, \ldots, u_{n-1}\}$ not adjacent to u_n.

Thus

$$d(u_n) \leq (n-1) - d(u_1).$$

Thus

$$d(u) + d(v) \leq n - 1.$$

This represents a contradiction. Thus G is Hamiltonian. ■

The concept discussed in Theorem 4.3 of adding an edge leads us to a useful definition concerning Hamiltonicity.

Definition

The *closure* of a graph G containing n vertices, denoted by $C(G)$, is the graph obtained from G by successively joining pairs of nonadjacent vertices whose degree sum is at least n, (in the graph obtained at each step of joining) until it is not possible to join any further pairs. This closure operation, which can be shown to be well defined, is illustrated in Figure 4.2. The next theorem is needed for a sufficiency condition for Hamiltonicity.

Theorem 4.4

A graph is Hamiltonian if and only if its closure is Hamiltonian.

Proof

By Theorem 4.3 and the definition of closure. ■

We come now to a sufficient condition for Hamiltonian graphs.

Theorem 4.5

Let G be a graph with at least 3 vertices. If $C(G)$ is complete, then G is Hamiltonian.

Proof

The proof is immediate by Theorem 4.4 and the fact that each complete graph with at least 3 vertices is Hamiltonian. ■

There are a number of corollaries of Theorem 4.5 which also provide sufficient conditions for Hamiltonicity, although they are weaker than Theorem 4.5 itself.

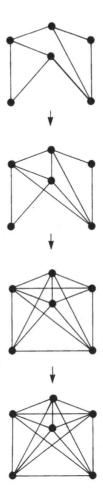

Figure 4.2 The closure operation

Corollary 4.1

If G is a graph with at least p vertices, where $p \geq 3$, such that for every integer i, with $1 \leq i \leq \frac{1}{2}p$, the number of vertices of degree not exceeding i is less than i, then G is Hamiltonian.

Corollary 4.2

If G is a graph with at least p vertices where $p \geq 3$, such that for all distinct nonadjacent vertices u and v,

$$d(u) + d(v) \geq p,$$

then G is Hamiltonian.

Corollary 4.3

If G is a graph with at least p vertices where $p \geq 3$, such that $d(v) \geq \frac{1}{2}p$ for every vertex v of G, then G is Hamiltonian.

In Section 4.2 we discussed first the quest for a closed trail in a graph, G, containing all of the edges of G exactly once. We then turned to finding an open trail containing all of the edges of G. Similarly, we now turn our attention to finding a path which covers all the vertices of a graph.

Definition

A spanning path in a graph is termed a *Hamiltonian path*.

Naturally, if a graph G, is Hamiltonian then it must contain Hamiltonian paths. These can be obtained by deleting exactly one edge from any Hamiltonian cycle. Conditions for a graph to contain a Hamiltonian path can be derived from the corollaries to Theorem 4.5.

It appears that there is little relationship between Eulerian and Hamiltonian graphs. An illustration of this is given in Figure 4.3, where (a) contains a graph which is both Eulerian and Hamiltonian, (b) contains a graph which is Eulerian but not Hamiltonian, (c) contains a graph which is Hamiltonian but not Eulerian, and finally (d) contains a graph which is neither Eulerian nor Hamiltonian.

4.4 Summary

We have now completed our discussion of the concept of traversability from the point of view of pure mathematics. We have covered two topics: that of a graph being Eulerian and secondly that of it being Hamiltonian. We have seen that these two concepts differ markedly in their degree of complexity. It is straightforward, via Theorem 4.1, to establish whether or not a graph contains an Euler trail. It is usually quite straightforward to actually identify an Eulerian trail, once it is known that one exists. The same cannot be said for Hamiltonian cycles. We shall discuss the application of these concepts in Chapter 12 on operations research.

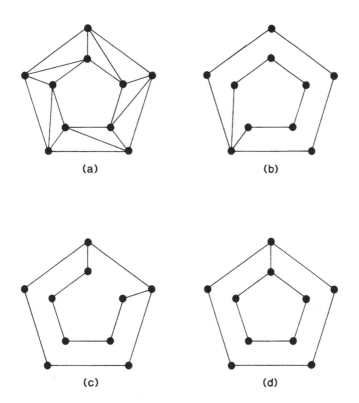

Figure 4.3 The lack of relationship between Eulerian and
Hamiltonian graphs

4.5 Exercises

4.1 You are given 10 dominoes with the following dots: (1,2), (1,6), (1,4), (1,5), (2,3), (2,4), (2,5), (3,4), (6,5), and (4,5). Using graph theory, discover if you can arrange the dominoes in a circle so that every pair of touching numbers are the same. Hint: Check a certain six-vertex graph to see if it is Eulerian. Repeat the exercise for the set: (1,2), (1,3), (4,5), (5,3), (4,3), (5,2), (4,2), (3,2), (5,1), and (4,1).

4.2 Repeat Exercise 4.1 where it is required to find a path rather than a circle.

4.3 With regard to Exercise 4.1, discuss in which sets of dominoes it is possible to find the required cycle, or the required path.

4.4 Is it possible to move a knight on a chessboard such that it touches each square exactly once and returns back to the original square from whence it started?

4.5 Repeat Exercise 4.4 where it is required to find merely a path rather than a cycle of moves.

4.6 Draw a diagram similar to Figure 4.3 which requires paths rather than cycles.

4.7 Prove that it is always possible to seat n people around a circular table such that each person has a friend on either side if each person is friendly with at least half of the others present.

4.8*Characterize graphs which are both Eulerian and Hamiltonian.

4.9*Characterize graphs which possess Hamiltonian paths but not Hamiltonian cycles.

4.10*Characterize graphs which are unicursal but not Eulerian.

4.11 Prove that a connected graph is unicursal if and only if it has exactly two vertices of odd degree. (Hint: See the proof of Theorem 4.2.)

4.12 Solve the following problem, by using a graph theoretic approach. You are given three jars: I, II, and III of capacity 8, 5, and 3 units respectively. I is filled, and II and III are empty. Divide the liquid in I so that 4 units of it remain in I. You may pour quantities from one vessel to another but cannot make any independent judgement about quantities.

4.13 Use the concept of Eulerian cycles to describe an algorithm for traversing a maze from a given starting point, to a given finishing point t, without traversing any corridor more than once in the same direction. Assume that the traverser has no memory and is allowed to use only two types of markers (say 0 and 1) to mark the entrances and exits of corridors.

5 Planarity

"Strike flat the thick rotundity o' the world!"

King Lear III, ii, William Shakespeare

In the previous chapters we have examined various properties of subgraphs. In this chapter we consider a property of a whole graph, which is of theoretical and practical importance. The concept revolves around the possibility of drawing a given graph in a plane such that its edges do not intersect. We shall find this property, termed *planarity,* very useful in Chapter 14. We begin with a simple puzzle in order to motivate the ideas to come.

5.1 The Utilities Problem

Suppose that there are three houses: h_1, h_2, and h_3, each of which is to be connected by cables to the centres of three companies which supply television, telephone service, and electricity. A schematic diagram indicating the cable service required is given in Figure 5.1(a). It is a requirement of all companies that the cables be laid underground in such a way that no cable crosses over the top of any other. The question is to find a layout of the cables so that each house can be supplied with the three services from the three centres in such a way that no two cables intersect. We can model this problem in graph theoretic terms by representing each of the

six locations by the vertex of a graph and each cable by an edge of the
graph directly joining the two vertices representing the locations which the
cable directly connects. A partial cable layout is given in Figure 5.1(b).
The reader will note, at least for the partial layout shown, that it is impossible to supply h_3 with television without cable intersection. Indeed, it
can easily be shown that any eight of the nine required cables can be laid
out without intersection, but that it is impossible to lay out all of the nine
in this way. We shall see that, in graph theoretic terms, this means that
the graph in Figure 5.1(a) is what we shall later call *nonplanar*. We now
make more rigorous the ideas of planarity that we have introduced loosely
via this utilities problem.

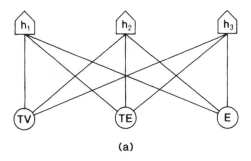

(a)

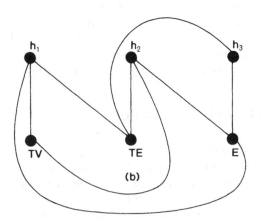

(b)

Figure 5.1 The utilities problem

5.2 Plane and Planar Graphs

Definitions

A graph G, is said to be *embedded* in a surface S, when it is drawn on S so that no two edges of G intersect.

A graph is said to be *planar* if it can be embedded in the plane.

A *plane graph* is a graph which is embedded in the plane.

A graph which cannot be embedded in the plane is said to be *nonplanar*.

As an example, any graph obtained from the graph in Figure 5.1(a) by the deletion of exactly one edge is planar because it is isomorphic to the plane graph in Figure 5.1(b).

Thus, returning to the utilities problem, we shall see that the graph of Figure 5.1(a) is nonplanar, making a cable layout without crossings impossible. We now ask the obvious question, how can we tell whether a given graph is planar or not? This is far from obvious for many graphs, especially when they contain many vertices and edges and are specified solely by their vertex and edge sets. One helpful fact is that the two graphs given in Figure 5.2 are both nonplanar. These are called the *Kuratowski graphs*, after the Polish mathematician Kasimir Kuratowski, who discovered some of their properties. The reader will recognise the first as the complete graph on five vertices K_5, and the second as the complete bipartite graph on six vertices, $K_{3,3}$, which is the graph of the utilities problem. We shall prove that both of these graphs are nonplanar later. It is interesting to compare the two of them. K_5 is the unique nonplanar graph with the smallest number of vertices. $K_{3,3}$ is the unique nonplanar graph with the smallest number of edges. Both are regular and each has the property that the removal of any edge makes it planar.

5.3 Planar Graph Representation

The following theorem is useful for the representation of graphs in the plane. It ensures that all of the edges of a plane representation of any graph can be drawn as straight line segments.

Theorem 5.1

There exists a plane representation of any planar graph with the property that every edge in the representation is a straight line segment.

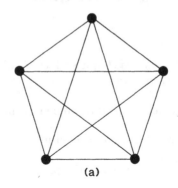

(a)

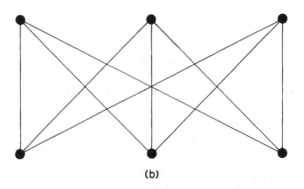

(b)

Figure 5.2 Kuratowski's graphs, K_5 and $K_{3:3}$

Proof

The reader is referred to the proof given by Fáry (1948).

Definitions

A plane representation of a graph divides the plane into areas which are termed *regions*. Regions are sometimes called *faces*, and are defined only for plane graphs.

The set of edges which bound a region of a plane graph is termed its *boundary*.

Of course, there will exist a region of infinite area in any plane graph G. This is the part of the plane which lies outside the plane representation of G. It is usually termed the *exterior region*. Regions of G which are not exterior are termed *interior*.

Theorem 5.2

Any planar graph may be embedded in a plane in such a way that any one of its regions may be made the exterior region.

Proof (Outline)

Consider a stereographic projection of a sphere S, on a plane P, accomplished as follows. Place S on P and let the unique point of contact between the two be denoted by a. Draw a straight line through a perpendicular to the plane. Denote the point (other than a) at which the line intersects the surface of the sphere by b. A one-to-one correspondence can be established between the points of S and the finite points of P as follows. Let p be a point on P. Then there is a unique point on S, corresponding to p, located at the intersection of the surface of S with the straight line joining b and p. Similarly, let s be a point on the surface of S (other than b). Then there is a unique point on P, corresponding to s, located at the intersection of P with the straight line joining b and s. If b is associated with the infinite points of P then a correspondence (not one-to-one) is established between all the points on the surface of S and of P. The one-to-one correspondence just explained, can be used to embed a plane graph G, in the surface of any sphere S. It will partition the surface of S into finite regions corresponding to the interior regions of G, with point b in a region corresponding to the exterior region of G. There always exists a rotation of the sphere which will associate the region on S containing b with the exterior region of G. ∎

The following theorem illustrates the relationship between the numbers of vertices, edges, and regions of any connected plane graph.

Theorem 5.3 (Euler's formula)

Any plane representation of a connected planar graph, with at least two regions, with n vertices and e edges has $e - n + 2$ regions.

Proof

Let G be a plane graph with n, e, and f vertices, edges, and regions, respectively. By Theorem 5.1 each region of G can be represented by a polygon. Let N_r be the number of regions of G which are bounded by r sides, and s be the number of sides of any polygon in G with the maximum number of sides. Then, because each edge of G bounds exactly two regions,

$$3N_3 + 4N_4 + \ldots + sN_s = 2e. \tag{5.1}$$

Further,

$$N_3 + N_4 + \ldots + N_s = f. \tag{5.2}$$

We now recall some elementary results from plane geometry. The sum of all the interior angles of any polygon with t sides is $\pi(t-2)$. The sum of all the exterior angles of any polygon with t sides is $\pi(t+2)$. Let us now calculate the sum of all angles subtended at each vertex of G. The part of the sum contributed by the $f-1$ interior regions is

$$\pi(3-2)N_3 + \pi(4-2)N_4 + \ldots + \pi(s-2)N_s.$$

The part of the sum contributed by the exterior region is 4π. Adding these together, and using (5.1) and (5.2), we obtain the total sum

$$2\pi(e - f + 2). \tag{5.3}$$

However the sum of all the angles subtended by the vertices of any polygonal net with n vertices is $2\pi n$. Equating this latter value with (5.3) produces the result. ∎

A number of simple corollaries follow immediately from Theorem 5.3.

Corollary 5.1

$$e \geq \tfrac{3}{2}f.$$

Corollary 5.2

$$e \leq 3n - 6.$$

We now use Theorem 5.3 and its corollaries to establish that K_5 and $K_{3,3}$ are both nonplanar.

Theorem 5.4

K_5 and $K_{3,3}$ are both nonplanar.

Proof

K_5 is nonplanar by Corollary 5.2.

It is clear from Figure 5.2(b) that each region of $K_{3,3}$ is bounded by at least four edges. Thus, if this graph is planar, the following inequality would hold:

$$2e \geq 4f.$$

Substituting $4f$ in the formula in Theorem 5.3:

$$2e \geq 4(e - n + 2).$$

But as $n = 6$ and $e = 9$, we obtain a contradiction. Hence the result follows. ■

Definition

A planar graph G is termed *maximally planar* if, for every pair of nonadjacent, distinct vertices u and v of G, the graph $G + uv$ is nonplanar.

This leads to the simple observation that any maximally planar graph has a 3-cycle as the boundary of each of its regions. Because 3-cycles are termed *triangles*, plane embeddings of maximally planar graphs are usually called *triangulations*. This leads to a special case of Corollary 5.2.

Theorem 5.5

If G is a triangulation with n vertices and e edges, where $n \geq 3$, then

$$e = 3n - 6.$$

Proof

Because the boundary of every region in G is a triangle, and each edge is on the boundary of two regions, the number of edges on the boundary of a region, summed over all regions equals $3f$. However such a sum includes each edge twice. Therefore

$$3f = 2e.$$

By Theorem 5.3 the result follows. ■

Theorem 5.6

Every planar graph contains a vertex of degree less than 6.

Proof

Let G be a planar graph with n vertices and e edges. If $n \leq 6$, then the result follows. Suppose $n > 6$. By Corollary 5.2, the degree sum is no more than $6n - 12$. Thus

$$2e \leq 6n - 12.$$

This implies that not all of the n vertices of G can have degree at least 6 because then

$$2e \geq 6n.$$

Hence the result follows. ■

5.4 Planarity Detection

As will be seen in Chapter 14, it is very important that we have an efficient way of establishing whether or not a given graph is planar. In administering a test of planarity for any graph, we can consider its components one at a time. For this reason we consider only connected graphs for the rest of this section. There is a further reduction that we can make that simplifies any planarity test. That is, we can eliminate any vertex of degree one or two from the graph under question by merging the two edges which are incident with it. If this is done progressively, eliminating all edges in series, the resulting graph will have the same planarity status as the original graph to be tested. Obviously, all but one of any set of parallel edges is removed should they arise due to the reduction just mentioned. This reduction process is illustrated in Figure 5.3.

Once the reduction just mentioned has been effected, the next obvious test is to employ Corollary 5.2 and to check whether the number of edges in the graph to be tested is less than or equal to $(3n - 6)$, where n is the number of vertices. In order to continue the test, should the graph still possibly be planar, we need the following concept.

Two graphs are said to be *homeomorphic* if one graph can be obtained from the other with a creation of edges in series (i.e. by the insertion of vertices degree 2) or by the merger of edges in series. Figure 5.4 illustrates two homeomorphic graphs.

The following two theorems are useful in planarity detection.

Theorem 5.7

A graph G is planar if and only if every graph that is homeomorphic to G is planar.

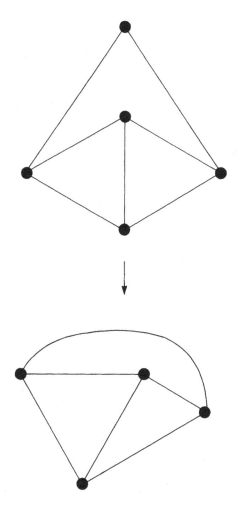

Figure 5.3 A reduction process to simplify
planarity testing

Proof

Obvious from the reduction process just stated. ∎

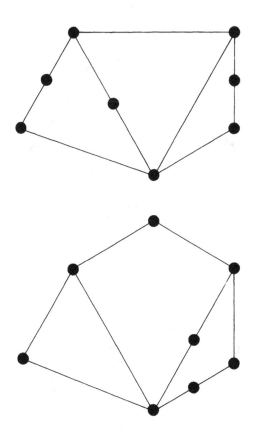

Figure 5.4 Two homeomorphic graphs

Theorem 5.8 (Kuratowski (1930).)

A graph G, is planar if and only if G does not contain a subgraph which is homeomorphic to either K_5 or $K_{3,3}$.

Proof

The proof is too involved to be given here. The reader is referred to Berge (1962, pp 211–213).

Note that Theorem 5.8 states that a sufficient condition for nonplanarity is that a graph possesses a subgraph which is homeomorphic to one of the Kuratowski graphs, but does not necessarily possess one of the Kuratowski graphs themselves. For example, neither of the nonplanar graphs in Figure 5.5 has either K_5 or $K_{3,3}$ as a subgraph. Although Theorem 5.5 does provide us with, at least in theory, a way of telling whether or not a given graph is planar, in practice it is difficult to implement it on all but trivial graphs. We now develop the concept of duality, which leads to more efficient planarity detection strategies.

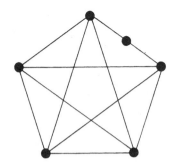

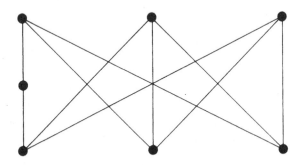

Figure 5.5 Two nonplanar graphs not containing K_5 or $K_{3,3}$ as subgraphs

5.5 Duality

We now construct what is known as the *geometric dual* of a given planar graph, in order to provide some intuitive motivation for the concept of duality in graph theory to be explained. Consider the pseudograph given in Figure 5.6(a). Vertices of another pseudograph to be constructed have been inserted in each of the regions of this plane pseudograph. We shall now construct a second pseudograph, with vertex set given by the square vertices in Figure 5.6(a), according to the following rules:

(i) Draw a vertex of the pseudograph to be constructed in each region of a plane representation of the given pseudograph.

(ii) Suppose two regions, say R_i and R_j, have an edge in common and contain new vertices v_i and v_j. Draw an edge connecting v_i and v_j that intersects only the common edge between R_i and R_j and intersects it only once. Repeat this process for each edge which is common between each R_i and R_j. Repeat this process for each pair of regions with an edge in common.

(iii) For each pendant edge which lies within region, say R_i, draw a loop incident with v_i around the new vertex associated with R_i. Ensure that the loop drawn intersects with the pendant edge exactly once and has no other intersections.

(iv) For each loop, draw an edge between the vertices in the two regions which have the loop in common.

Application of the above rules produces the pseudograph in Figure 5.6(b), which has square vertices and dotted edges. For clarity it is reproduced in Figure 5.6(c). If G' is a pseudograph produced by the application of the above rules from a plane pseudograph G, then G' is called a *geometric dual* of G.

On studying Figure 5.6(b) we can see that the edges of the pseudograph and its dual are paired in the sense that there is exactly one crossing between each pair. This always happens when the dual of a pseudograph is constructed, and thus a pseudograph and its dual always have the same number of edges. Before stating some properties of pseudograph geometric duality, we introduce some concepts that we shall find useful here and also later in this book.

Let G be a pseudograph with n vertices, e edges, and k components.

Definitions

The *rank r* of G is

$$r = n - k.$$

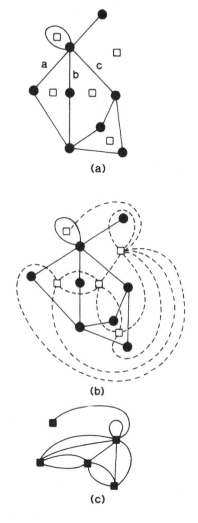

Figure 5.6 A pseudograph and its
geometric dual

The *cyclomatic number* μ of G is

$$\mu = e - n + k.$$

If G is connected, its rank is $(n-1)$ and its cyclomatic number is $(e-n+1)$.

There are a number of trivial deductions that we can make concerning these two concepts for any graph G:

   (i) r = the number of branches in any spanning forest of G.

   (ii) μ = the number of chords in G with respect to any spanning forest of G.

   (iii) $r + \mu = e$.

We return to make some observations about the relationship between a planar pseudograph G, with n vertices, e edges and f regions and the pseudograph which is its geometric dual G', with n' vertices, e' edges, and f' regions.

   (i) If G and G' have rank and cyclomatic number: r, μ and r' and μ', respectively, then:

   $n' = f$

   $e' = e$, (as observed earlier),

   $f' = n$,

   $r' = \mu$,

   $\mu' = r$.

   (ii) Suppose that v_i is a vertex in G associated with region R' in G'. Then the boundary of R' has $d(v_i)$ edges.

   The last two observations lead to a number of simple results. Suppose that e is an edge in G which is matched by intersection in the construction of its dual with edge e' of G.

   (iii) If e is a loop then e' is pendant.

   (iv) If e is pendant then e' is a loop.

   (v) Edges belonging to a path in G correspond to parallel edges in G'.

   (vi) Parallel edges in G correspond to a path in G'.

   (vii) G' is planar.

By reversing the sequence of diagrams in Figure 5.6, it is clear that the dual of the dual of the original graph G, is G itself. It is natural to ask whether or not this is true in general. In other words, is the dual of the dual of any planar graph G, isomorphic to G? The answer is *yes* if G is connected.

We can ask a related question: Are all duals of a planar pseudograph isomorphic? The key to the answer lies in the realization that a planar pseudograph will have a unique dual (up to isomorphism) if its embedding in the plane is unique. As any 3-connected graph can be shown to have a unique embedding in the sphere, it can be shown that all of its duals are isomorphic. But what of graphs which are not 3-connected? As an example, Figure 5.7 (a) and (b) show two distinct embeddings of a graph

G. The duals of the embeddings of (a) and (b) are shown in Figure 5.7 (c) and (d) respectively. Although the duals in (c) and (d) are not isomorphic, they are 2-isomorphic. This is true in general, as stated in the following theorem.

Theorem 5.9

Every graph which is 2-isomorphic to a dual of a planar graph G, is a dual of G. If graphs G_1 and G_2, are duals of G then G_1 and G_2 are 2-isomorphic.

Proof

A proof can be found in Deo (1974).

We now develop the notion of a *combinatorial dual*, which is independent of the geometric notions of duality that we found so useful in analysing Figure 5.6. In order to achieve this analytic notion of duality we must first introduce the concept of a *cut-set*.

Definition

A *cut-set* in a connected graph $G = (V, E)$, is a minimal set of edges of E whose removal from G, renders G disconnected.

The edge set $\{a, b, c\}$, constitutes a cut-set of the graph in Figure 5.6(a). We can now state an analytic characterization of duality.

Theorem 5.10

A necessary and sufficient condition for two planar pseudographs: G and G', to be the duals of each other is: there exists a one-to-one correspondence of the edges in G with the edges in G' such that a subset of the edges in G forms a cycle if and only if the corresponding set of edges in G' forms a cut-set. (When such a correspondence exists, G' is called a *combinatorial dual* of G.)

Proof

(Necessity)
Let $G = (V, E)$, be a planar pseudograph containing a cycle C. Suppose that a dual $G' = (V', E')$, of G is constructed from a plane embedding of G by the process illustrated in Figure 5.6. By the Jordan Curve Theorem, C will partition the plane in which G is embedded into two connected regions which are both homeomorphic to a disk. Also C partitions V into two nonempty subsets say V_1' and V_2', one inside C and one outside C. Let the set of edges in G' corresponding to the edges of C be denoted by C'. It

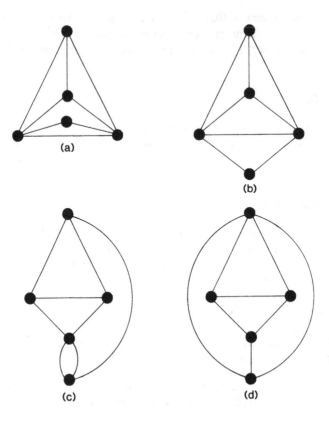

Figure 5.7 Two embeddings of a planar graph with
2–isomorphic duals

is easily seen that C' constitutes a cut-set of G'. This is so because the removal of C' disconnects G' (by disconnecting V_1' from V_2') in G', and no proper subset of C' disconnects G'. Let S' be any cut-set of G', and let S be the set of edges in G corresponding to them according to the process illustrated in Figure 5.6. Suppose that S' partitions V' into V_1' and V_2'. Then S is the unique set of edges of G which partition the plane in which G is embedded into two regions, one containing V_1' and one containing V_2'. Thus S is the unique cycle in G whose edges correspond to the edges of S' in G'.

(Sufficiency)
Let $G = (V, E)$, be a planar pseudograph and $G' = (V', E')$, be a pseudo-graph such that there is a one-to-one correspondence between the edges of

E and E' such that any subset of E constituting a cycle in G corresponds to a cut-set in G', and any subset of E' constituting a cut-set in G' corresponds to a cycle in G. Thus there is a one-to-one correspondence between the cut-sets of G and the cycles of G'. There is also a one-to-one correspondence between the subsets of G and the cycles of any dual G^1, say, of G. Therefore there is a one-to-one correspondence between the cycles of G' and the cycles of G^1. Thus G' and G^1 are 2-isomorphic.

Thus G' is a dual of G. ■

Lemma 5.1

Neither K_5 nor $K_{3,3}$ has a combinatorial dual.

Proof

By contradiction.

(i) Suppose that K_5 has a dual, say K'. On referring to Figure 5.2(a) it can be seen that K' must have ten edges, all vertices of degree at least four, no parallel edges, and each of its cycles is either a 4-cycle or a 6-cycle with at least one 6-cycle, say C. However, four or more edges cannot be added to C without creating either: (i) a 3-cycle or (ii) a pair of parallel edges. Further K' cannot possess either of these. As K' has ten edges, it must have at least seven vertices. Because every vertex in K' has degree at least three, K' must have at least eleven edges which is a contradiction.

(ii) Suppose that $K_{3,3}$ has a dual K''. By Theorem 5.10, the cycles (cut-sets) of $K_{3,3}$ correspond to the cut-sets (cycles) of K''. However $K_{3,3}$ does not possess a cut-set of two edges. Thus K'' does not possess a pair of parallel edges. Also, every cycle in $K_{3,3}$ is either a 4-cycle or a 6-cycle. Thus every cut-set in K'' has more than three edges. Thus every vertex in K'' has more than four vertices, each with degree more than three. Hence K'' has at least ten edges. But this is impossible because $K_{3,3}$ has only nine edges. ■

Theorem 5.10 allows us to ask the question, do all pseudographs have duals? Duals of nonplanar pseudographs cannot be constructed by the geometric process illustrated in Figure 5.6. However, the question is valid given Theorem 5.10, which is divorced from geometric notions. The question can be rephrased as follows. Suppose that G is an arbitrary nonplanar pseudograph. Is it possible to identify another pseudograph G' such that the edges of G and G' are paired in a one-to-one fashion so that every subset of edges in G comprising a cycle corresponds to a unique set of edges in G' which represents a cut-set in G', and every cut-set in G corresponds to a unique cycle in G'? Theorem 5.11 can be interpreted as providing the answer to our original question, in the negative. That is, nonplanar pseudographs do not have duals.

Theorem 5.11

A pseudograph possesses a dual if and only if it is planar.

Proof (Outline)

Let G be a graph. It is clear that if G is planar then G has a dual. To prove the converse, assume that G is nonplanar. We shall show that G does not have a dual. By Theorem 5.8, G contains a subgraph homeomorphic to either K_5 or $K_{3,3}$. It can be shown that a graph can have a dual only if every subgraph of it and every graph homeomorphic to the subgraph has a dual. However by Lemma 5.1, neither K_5, nor $K_{3,3}$ has a dual. Hence the result follows. ∎

Definition

A pseudograph G is said to be *self-dual* if G is isomorphic to its dual.

Along with Theorem 5.8, Theorem 5.11 provides us with a second way of testing to see whether or not a given graph is planar. Unfortunately neither approach has led to a planarity detection algorithm that is easy to implement on large, arbitrary graphs. However there does exist a graph planarity detection algorithm which is efficient in the sense that the number of elementary computational steps required to employ it grows linearly with the number of vertices. The efficiency of graph theoretic algorithms is discussed in Section 9.3 along with an outline of the method in Section 9.5. An application of the method, due to Hopcroft and Tarjan (1974), will be discussed in Section 14.2.

5.6 Thickness and Crossing Numbers

By definition, if a graph is planar, it can be embedded in a single plane. Suppose however, that we are given a nonplanar graph. How many planes must be required in order to fully embed it? This idea leads to the following definition.

Definition

The least number of planar subgraphs whose union is a given graph G, is termed the *thickness* of G.

This means that the thickness of any planar graph, by definition, is 1. Also K_5 and $K_{3,3}$ have thickness 2. However the thickness of K_9 is 3.

The discussion on the utilities problem, illustrated in Figure 5.1, leads one to ask what is the fewest number of edge intersections which are necessary in order to create a geometric representation of a given graph in the plane? Note here that we are not necessarily suggesting an embedding of any given graph, as the graph may well be nonplanar and thus require intersection of the line segments used to represent its edges.

Definition

The minimum number of edge intersections required in order to represent a graph G, in the plane is said to be the *crossing number* of the graph.

Thus the crossing number of any planar graph is 0 and of K_5 and $K_{3,3}$ is 1.

The search for formulas for minimal crossing numbers was initiated in 1944 by the Hungarian mathematician P. Turan. He was working in a war labour camp, at a brick factory outside Budapest. Turan (1977) recalls:

"There were some kilns where the bricks were made and some open storage yards where the bricks were stored. All the kilns were connected by rail with all the storage yards. The bricks were carried on small wheeled trucks to the storage yards. All we had to do was to put the bricks on the trucks at the kilns, push the trucks to the storage yards, and unload them there. We had a reasonable piece rate for the trucks, and the work itself was not difficult: the trouble was only the crossings. The trucks generally jumped the rails there, and the bricks fell out of them. In short, this caused a lot of trouble and loss of time which was rather precious to all of us (for reasons not to be discussed here). We were all sweating and cursing at such occasions, I too; but **nolens-volens** *the idea occurred to me that this loss of time could have been minimized if the number of crossings of the rails had been minimized. But what is the minimum number of crossings? I realized after several days that the actual situation could have been improved, but the exact solution of the general problem with m kilns and n storage yards seemed to be very difficult and again I postponed my study of it to times when my fears for my family would end. The problem occurred to me again not earlier than 1952, at my first visit to Poland, where I met Zarankiewicz. I mentioned to him my "brick-factory"-problem."*

The Polish mathematician S. Zarankiewicz believed he had solved the graph crossing-number problem, but a gap was found in his proof, and the problem became the notorious unsolved question that it remains today. We state only a few specialized results concerning thickness.

Theorem 5.12

If θ_n is the thickness of K_n then

$$\theta_n = \begin{cases} [(n+7)/6], & n \neq 9,10 \\ 3, & n = 9,10. \end{cases}$$

Proof

The reader is referred to the proof given by Beineke (1967).

Theorem 5.13

If ν_n is the crossing number of K_n then

$$\nu_n \leq \tfrac{1}{4}[n/2][(n-1)/2][(n-3)/2],$$

with equality for $n \leq 10$.

Proof

The reader is referred to the proof given by Guy (1960).

Theorem 5.14

If $\nu_{m,n}$ is the crossing number of $K_{m,n}$ then

$$\nu_{m,n} \leq [m/2][(m-1)/2][n/2][(n-1)/2],$$

with equality for $m \leq 6$.

Proof

The reader is referred to the proof given by Guy (1972).

Corollary

If $m \geq 5$ and $n \geq 7$ then

$$\nu_{m,n} \geq [m(m-1)/20][(n-1)(11n-5)/12].$$

5.7 Summary

In this chapter we first introduced the concept of graph planarity via a trivial puzzle and then discussed questions of planar graph representation, and how to test for its existence. One of the cornerstones of this area of graph theory is Kuratowski's theorem.

We introduced the concept of duality in graph planarity in order to provide a second characterization of planar graphs. We ended the chapter with a brief discussion of thickness and crossing numbers. These ideas have obvious application in printed circuit board design, and will be discussed in Section 13.2.

5.8 Exercises

5.1 Prove Theorem 5.4 using solely geometric arguments. Note that a graph can be embedded in the surface of a sphere if and only if it can be embedded in a plane.

5.2 Prove that any planar graph can be embedded in a plane in such that a way that any of its regions can be made the infinite region.

5.3 What are necessary and sufficient conditions for the geometric dual of a graph to be unique?

5.4*Characterize complete, self-dual graphs.

5.5 Establish the crossing number of K_6.

5.6 A planar graph G is said to be *completely regular* if the degrees of all vertices of G are equal and every region is bounded by the same number of edges. Show that there are only five possible simple completely regular planar graphs with vertex degree > 2. Hint: use Euler's formula.

5.7 Draw the geometric duals of the graphs in Figures 5.3 and 5.4. Prove the observations given in Section 5.5.

5.8 If every region of a simple planar graph (with n vertices and e edges) embedded in a plane is bounded by k edges, show that $e = k(n - 2)/(k - 2)$.

5.9 Show that a planar graph with at least four vertices has at least four vertices with degree five or less.

5.10 Prove that the edges forming a spanning a tree in a planar graph G correspond to the edges forming a set of chords in the dual of G.

5.11 Prove that the complete graph on four vertices is self-dual. Give another example of a self-dual graph.

6 Matrices

"Now go, write it before them in a table, and note it in a book."

Isaiah, xxx, 8, The Bible

A graph is completely determined by specifying either its adjacency structure or its incidence structure. These specifications provide far more efficient ways of representing a large or complicated graph than a pictorial representation. Because computers are more adept at manipulating numbers than at recognizing pictures, it is standard practice to communicate the specification of a graph to a computer in matrix form.

There are a number of matrices which we can associate with any graph which has both its vertices and its edges labelled: the *adjacency matrix*, the *incidence matrix*, the *cycle matrix*, the *cut-set matrix*, and the *path matrix*. Not only are these matrices useful devices for storing the basic structure of any graph, they can be manipulated in order to study its properties. This turns out to be especially useful in electrical engineering and operations research. Ideas will be developed, along with the corresponding digraph matrices in Chapter 7, and will be applied later in this book in our analysis of electrical networks in Chapter 13. Throughout this chapter we shall deal with a graph G, with n vertices, e edges, and k components. Unless otherwise stated, we shall further assume that $k = 1$.

6.1 The Adjacency Matrix

Definition

The *adjacency matrix* $A = (a_{ij})_{n \times n}$, of a vertex-labelled graph G, with n vertices, is the matrix in which $a_{ij} = 1$ if vertex v_i is adjacent to v_j in G and $a_{ij} = 0$ otherwise, where v_i and v_j are vertices of G.

A diagram of a labelled graph and its adjacency matrix are shown in Figure 6.1(a). There are a number of observations that we can make about the adjacency matrix A, of a labelled graph G.

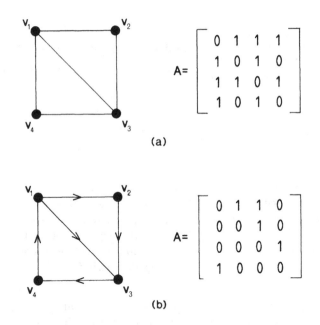

(a)

(b)

Figure 6.1 Adjacency matrices of a graph and a digraph

A(i) A is symmetric.

A(ii)   The sum of the entries in each row i of A equals the degree of v_i.

A(iii)   There is a one-to-one correspondence between labelled graphs with n vertices and $n \times n$ symmetric binary matrices with all entries on the leading diagonal equal to zero.

A(iv)   G is connected if and only if there is no labelling of the vertices of G such that its adjacency matrix is a block diagonal matrix.

A(v)   If A_1 and A_2 are adjacency matrices which correspond to different labellings of the same graph G, then for some permutation matrix P,

$$A_1 = P^{-1} A_2 P.$$

A(vi)   The (i, j) entry of A^m is the number of walks of length m from vertex v_i to vertex v_j, in G. This observation has a number of consequences:

A(vii)   If $i \neq j$, the (i, j) entry of A^2 is equal to the number of paths containing exactly two edges from v_i to v_j. The (i, i) entry of A^2 is the degree of v_i and that of A^3 is equal to twice the number of triangles containing v_i.

A(viii)   If G is connected, the distance between its vertices v_i and v_j, for $i \neq j$, is the least integer m, for which the (i, j) entry of A^m is nonzero.

Definition

The *adjacency matrix* of a vertex-labelled digraph D, with n vertices is the matrix $A = (a_{ij})_{n \times n}$, in which $a_{ij} = 1$ if arc $v_i v_j$ is in D, and 0 otherwise.

We can make a number of observations about the adjacency matrix of a digraph:

DA(i)   A is not necessarily symmetric.

DA(ii)   The sum any of column j of A is equal to the number of arcs directed towards v_j.

DA(iii)   The sum of the entries in row i is equal to the number of arcs directed away from vertex v_i.

DA(iv)   The (i, j) entry of A^m is equal to the number of walks of length m from vertex v_i to vertex v_j.

A digraph and its adjacency matrix are shown in Figure 6.1(b).

6.2 The Incidence Matrix

Consider a graph G, which has n vertices, and m edges, all labelled.

Definition

The *incidence matrix* of a vertex- and edge-labelled graph G, is the matrix $B = (b_{ij})_{n \times m}$, in which $b_{ij} = 1$ if vertex v_i and edge e_j are incident in G and $b_{ij} = 0$ otherwise.

A graph and its incidence matrix are shown in Figure 6.2(a). We can make a number of observations about the incidence matrix B, of a graph G.

I(i)   There is a one-to-one correspondence between vertex- and edge-labelled graphs and binary matrices which have exactly two unit entries in each column.

I(ii)   Any $n - 1$ rows of B determine its corresponding graph. This is so because each row of B is the sum of all other rows of B (modulo 2).

I(iii)   Each column of B comprises exactly two unit entries.

I(iv)   A row with all zero entries corresponds to an isolated vertex.

I(v)   A row with a single unit entry corresponds to a pendant vertex.

I(vi)   The number of unit entries in row i of B is equal to the degree of the corresponding vertex v_i.

I(vii)   The permutation of any two rows (any two columns) of B corresponds to a relabelling of the vertices (the edges) of G.

I(viii)   Two graphs are isomorphic if and only if their corresponding incidence matrices differ only by a permutation of rows or columns.

I(ix)   If G is connected with n vertices then the *rank* of B is $n - 1$.

The reader should compare this observation with the definition of *rank*, given in Section 5.5. This observation can be generalized:

I(x)   If G has k components, then the rank of B is $n - k$.

Because the rows of B are linearly dependant, it is natural to analyze the submatrix of B created by the removal of any one row. Such a matrix is called the *reduced incidence matrix*, and is denoted by B_r.

I(xi)   If G is connected, the rows of B_r are linearly independent.

I(xii)   If G is a tree with n vertices, then B_r is nonsingular and square, with order and rank $(n - 1)$.

We now define the incidence matrix of a digraph.

Definition

The *incidence matrix* $B = (b_{ij})_{n \times m}$, of vertex- and arc-labelled digraph D, with n vertices and m labelled arcs, is the $n \times m$ matrix in which $b_{ij} = 1$ if arc j is directed away from a vertex v_i, $b_{ij} = -1$ if arc j is directed towards vertex v_i, and $b_{ij} = 0$ otherwise.

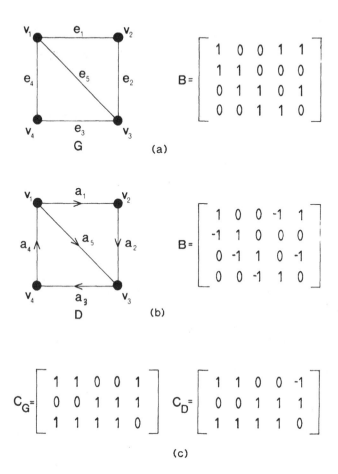

Figure 6.2 Incidence and Cycle matrices

The reader is encouraged to make analogous observations for the incidence matrix of a digraph as have been made for the incidence matrix of a graph above.

6.3 The Cycle Matrix

Definition

The *cycle matrix* $C = (c_{ij})_{c \times m}$, of an edge- and cycle-labelled graph G, is the matrix in which $c_{ij} = 1$, if the i-th cycle of G contains the edge e_j and $c_{ij} = 0$ otherwise.

The cycle matrix for the graph shown in Figure 6.2(a) is given in Figure 6.2(c).

The cycles of the graph G in Figure 6.2(a) are:

$$Z_1 = (e_1, e_2, e_5),$$
$$Z_2 = (e_3, e_4, e_5), \qquad \text{and}$$
$$Z_3 = (e_1, e_2, e_3, e_4).$$

We can make a number of observations concerning the cycle matrix C, of a graph G.

C(i) C does not determine G up to isomorphism. That is, it is possible for two nonisomorphic graphs to share the same cycle matrix. This observation still holds even when edges which do not belong to a cycle are excluded.

C(ii)   A column of all zeros in C corresponds to an edge which does not belong to any cycle.

C(iii)   The number of unit entries in any row corresponds to the number of edges in the cycle that the row represents.

C(iv)   Permutation of the rows (the columns) in C corresponds to a relabelling of the cycles (the edges) of G.

Let M^T denote the transpose of any matrix M.

Theorem 6.1

If B is the incidence matrix and C is the cycle matrix of a graph in which the columns have been arranged using the same order of edges then

$$BC^T = 0,$$

and

$$CB^T = 0 \pmod 2.$$

Proof

The p-th entries in the i-th row of C and the j-th column of B^T (which is the j-th row of B) are both nonzero if and only if edge e_p, is in the i-th cycle Z_i, and is incident with vertex v_j. If e_p is in Z_i, then v_j is also, but if v_j is in the cycle Z_i, then there are two edges of Z_i, incident with v_j so that the (i, j) entry of CB^T equals $1 + 1 = 0 \pmod 2$. ■

Let $D = (V, A)$, be a digraph whose arcs have been labelled and whose cycles have been both labelled and given an arbitrary orientation.

Definition

The *cycle matrix* $C = (c_{ij})$, of a digraph D, is a matrix in which $c_{ij} = 1$ if cycle Z_i of D contains arc a_j directed in the same way as the orientation of Z_i, $c_{ij} = -1$ if Z_i contains arc a_j directed in the opposite way to the orientation of Z_i, and $c_{ij} = 0$ otherwise.

We discussed the concept of a set of fundamental cycles in Section 3.7. The set of fundamental cycles, which are constructed with regard to a spanning tree in a connected graph, are independent, and also, representations of all of the other cycles of the graph can be obtained as linear combinations of the rows representing the fundamental cycles in the cycle matrix.

Definition

A submatrix of a cycle matrix of a graph G, which has a set S, of fundamental cycles, in which each row represents a fundamental cycle of S, is called a *fundamental cycle matrix*, and is denoted by C_f.

A spanning tree, comprising the edges: e_1, e_4, and e_5, of the graph in Figure 6.2(a) has been shown in Figure 6.3(a). With respect to this spanning tree, there are two fundamental cycles: $\langle e_1, e_2, e_5 \rangle$, and $\langle e_3, e_4, e_5 \rangle$. The corresponding fundamental cycle matrix C_f, is also shown in Figure 6.3(a).

If G is a graph with n vertices and e edges, and is connected, then C_f is of dimension $(e - n + 1) \times e$. This is because any spanning tree of G has $n - 1$ branches and hence $e - n + 1$ chords. This means that, because each chord gives rise to exactly one fundamental cycle, C_f will have $e - n + 1$ rows.

Figure 6.3(b) shows a permutation of the columns of the fundamental cycle matrix in Figure 6.3(a) in which the submatrix of C_f, corresponding to the columns which represent the chords of G, is an identity matrix.

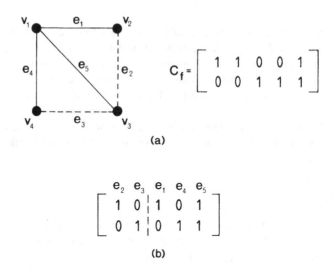

(a)

$$\begin{array}{ccccc} e_2 & e_3 & e_1 & e_4 & e_5 \end{array}$$
$$\begin{bmatrix} 1 & 0 & 1 & 0 & 1 \\ 0 & 1 & 0 & 1 & 1 \end{bmatrix}$$

(b)

$$K = \begin{bmatrix} 1 & 0 & 0 & 1 & 1 \\ 1 & 1 & 0 & 0 & 0 \\ 0 & 1 & 1 & 0 & 1 \\ 0 & 0 & 1 & 1 & 0 \\ 0 & 1 & 0 & 1 & 1 \\ 1 & 0 & 1 & 0 & 1 \end{bmatrix}$$

(c)

$$K_f = \begin{bmatrix} 1 & 1 & 0 & 0 & 0 \\ 0 & 0 & 1 & 1 & 0 \\ 0 & 1 & 1 & 0 & 1 \end{bmatrix}$$

(d)

Figure 6.3 (a) A graph and its fundamental cycle matrix C_f,
(b) a permutation of C_f, (c) its cut-set matrix K and
(d) a fundamental cut-set matrix K_f

Such a permutation of C_f for any connected graph can, of course, always be achieved. Thus C_f for any connected graph can be rearranged through chord column permutation to be expressed as

$$C_f = (I_\mu : C_t),$$

where I_μ is an identity matrix of dimension $\mu = e - n + 1$, and C_t is the remaining matrix of dimension $\mu \times n - 1$, corresponding to the branches of the spanning tree. It is now evident why $\mu = e - n + 1$ was defined to be the rank of the corresponding graph. This is because the rank of $C_f = \mu$. Because C_f is a submatrix of C, the rank of C is no less than $e - n + 1$. Indeed, we can state a sharper result.

Theorem 6.2

If G is a connected graph with e edges and n vertices then the rank of C, its cycle matrix, is $(e - n + 1)$.

Proof

Let B be the incidence matrix of G. Let $r(B)$ and $r(C)$ denote the ranks of B and C respectively. By Theorem 6.1 we have that

$$CB^T = 0 \quad (\text{mod } 2).$$

Via a well-known result in matrix theory it follows that:

$$r(C) + r(B) \le e.$$

We have, by observation I(ix), that

$$r(C) \le e - n + 1.$$

But the rank of the submatrix C_f, of C is

$$\mu = e - n + 1.$$

Hence

$$r(C) \ge e - n + 1.$$

Thus the result follows. ∎

This result can be generalized:

Theorem 6.3

If G is a graph with n vertices, e edges, and k components with cycle matrix C, then
$$r(C) = \mu = e - n + k.$$

Proof

By applying Theorem 6.2 to each component of G. ∎

Note that the two rows of C_f, representing the two fundamental cycles of the graph in Figure 6.3(a), can be added (mod 2) together to produce the third cycle (which is not fundamental), represented by the third row of the cycle matrix in Figure 6.2(c). We can perform a similar exercise to that just done for C_f, to digraphs. In this case each fundamental cycle is oriented according to the direction of the chord which gives rise to it. As an exercise, the reader should create the fundamental cycle matrix for the digraph in Figure 6.2(b) with respect to the two chords a_2, and a_3.

6.4 The Cut-Set Matrix

We introduced the notion of a *cut-set* in Section 5.5. Let G be a graph whose cut-sets and edges have been labelled.

Definition

The *cut-set matrix* $K = (k_{ij})$, of a graph G is a matrix in which $k_{ij} = 1$ if the i-th cut-set of G contains the edge e_j, and $k_{ij} = 0$ otherwise.

The cut-set matrix for the graph in Figure 6.3(a) is shown in Figure 6.3(c). For a planar graph, the problem of counting the number of its cut-sets is equivalent to counting the number of cycles in its dual. We now state a result which relates the rank of two different matrices of a connected graph.

Theorem 6.4

If C is the cycle matrix and K is the cut-set matrix of a graph G, in which the columns have been arranged using the same order of edges, then

$$CK^T = 0,$$

and

$$KC^T = 0 \quad (\text{mod } 2).$$

Proof

The number of edges common to any cut-set and any cycle of G is even (See Exercise 6.1). Thus every row of K is orthogonal (mod 2) to every row of G, assuming the same arrangement of columns of K and of C. ∎

As with cycle matrices, the rows of K are usually linearly dependent. Once again it is useful to develop a fundamental version of the matrix. Let T be a spanning tree of G. Let $b = v_1 v_2$ be any branch of T. Because the set $\{b\}$ is a cut-set of T, it partitions V into two disjoint sets (say V_1, and V_2), one containing v_1, and one containing v_2. Consider the cut-set K_f, containing only one branch, namely b, and any chords in G with respect to T, necessary to disconnect V_1 and V_2. A cut-set constructed in this way is termed a *fundamental cut-set*. It contains exactly one branch of T and is defined with respect to T.

Definition

The *fundamental cut-set matrix* of a graph G, is matrix $K_f = (k_{ij})_{(n-1)\times e}$ in which $k_{ij} = 1$, if the i-th fundamental cut-set in G contains the j-th edge and $k_{ij} = 0$ otherwise.

Corollary 6.1

$$C_f K_f^T = 0,$$

and

$$K_f C_f^T = 0 \quad (\text{mod } 2).$$

Theorem 6.5

If G is a connected graph with n vertices, e edges, rank r, cut-set matrix K, and incidence matrix B, then

$$r = r(K) = r(B).$$

Proof

For arbitrary graphs:
$$r(K) \geq r(B).$$

Because G is connected,
$$r(K) \geq n - 1. \tag{6.1}$$

By Theorem 6.4,
$$r(C) + R(K) \leq e.$$

However as G is connected,
$$r(C) = e - n + 1,$$

and

$$r(K) \leq n - 1. \tag{6.2}$$

From (6.1) and (6.2) the result follows. ∎

We now illustrate these notions via the graph in Figure 6.3(a), Let T have branch set $\{e_1, e_4, e_5\}$. Removal of branch $b = e_1$, partitions V into $V_1 = \{v_1, v_3, v_4\}$ and $V_2 = \{v_2\}$. The only chord connecting V_1 and V_2 is e_2. Thus we have a fundamental cut-set $\{e_1, e_2\}$. The complete fundamental cut-set matrix is given in Figure 6.3(d).

With respect to a given spanning tree T, in a connected graph, each branch of T gives rise to a *unique* fundamental cut-set. We now develop the relationship between fundamental cycles and fundamental cut-sets.

Theorem 6.6

Let d be a chord with respect to a given spanning tree T, in a connected graph G. Suppose that d determines a fundamental cycle C, say, in G. Then d belongs to all of the fundamental cut-sets created by the branches of C and does not belong to any other fundamental cut-set of G.

Proof

Let d_i be a chord with respect to a given spanning tree T in a connected graph G.

Suppose that the fundamental cycle created by d_i, with respect to T, is

$$c_f = \langle d_i, b_1, b_2, \ldots, b_p \rangle,$$

where b_j is a branch of T, $j = 1, 2, \ldots, p$.

Suppose that the fundamental cut-set created by b_j, where $b_j \in C_f$, with respect to T, is

$$k_f = \{b_j, d_1, d_2, \ldots, d_q\},$$

where d_h is a chord with respect to T, $h = 1, 2, \ldots, q$.

Because every cycle has an even number of edges in common with any cut-set (see Exercise 6.1), c_f and k_f have an even number of edges in common. As b_j is the only branch in both c_f and k_f, there must be at least one chord in both sets. As d_i is the only chord in c_f it must also be in k_f. Thus d_i is one of the chords of k_f. This argument can be repeated for all b_j, $j = 1, 2, \ldots, p$. Thus d_i belongs to every fundamental cut-set created by the branches of T. However d_i does not belong to any other fundamental cut-set, say k'_f. This is because none of the branches of c_f are in k'_f. Thus only edge d_i would be in both k'_f and c_f. This contradicts the fact that every cycle has an even number of edges in common with every cut-set. ■

Theorem 6.6 can be illustrated via the graph and its matrices C_f and K_f, in Figure 6.3. Consider the spanning tree T, with edges: e_1, e_4, and e_5, and chords: e_2 and e_3. Let $d_i = e_2$ be the chord chosen with respect to T, which creates the fundamental cycle

$$c_f = \langle e_2, e_5, e_1 \rangle,$$

corresponding to the first row of C_f in Figure 6.3(a).

The fundamental cut-sets created by the branches of C_f are:

$$b_j = e_5 : k_f = \{e_5, e_2, e_3\}, \qquad \text{and}$$
$$b_j = e_1 : k_f = \{e_1, e_2\}.$$

These equations correspond to the first and second rows of K_f in Figure 6.3(d). Chord e_2 occurs in each of these fundamental cut-sets but not in the only other fundamental cut-set, as evidenced by the appearance of a zero in the $(2, 2)$ entry of K_f.

Theorem 6.7

If b is a branch with respect to a spanning tree in a connected graph G that determines a fundamental cut-set k_f, then b is a member of every fundamental cycle created by the chords of k_f and does not belong to any other fundamental cycle.

Proof

By analogous arguments to those used in the proof of Theorem 6.6. ■

To illustrate Theorem 6.7 we shall use the graph and its matrices shown in Figure 6.3. Let $b = e_5$ be the branch of the spanning tree $T = \{e_1, e_4, e_5\}$. Branch e_5, determines the fundamental cut-set $k_f = \{e_5, e_2, e_3\}$, represented by the second row of K_f in Figure 6.3(d).

The two fundamental cycles created by the chords of k_f are:

$$b_i = e_2 : c_f = \langle e_2, e_5, e_1 \rangle, \qquad \text{and}$$
$$b_i = e_3 : c_f = \langle e_3, e_4, e_5 \rangle,$$

corresponding to the two rows of C_f in Figure 6.3(a). Branch e_5 is contained in both of these fundamental cycles.

Consider a connected graph G, with rank μ, and spanning tree T, which gives rise to

$$C_f = [I_\mu : C_t], \qquad \text{and}$$
$$K_f = [K_d : I_{n-1}],$$

where the edges have been given the same order in both of the above equations. Here d is the subscript associated with the chords with respect to T. Also, t is the subscript associated with the branches of T. The *reduced incidence* matrix B_r, of G, (created by removing one row from B) has the same edge ordering as above. We partition B_r as:

$$B_r = [B_d : B_t],$$

where B_t comprises the $n - 1$ columns representing the branches of T and B_d comprises the remaining $e - n + 1$ columns representing the chords with respect to T.

By Theorem 6.1, we have

$$B_r C_f^T = 0 \quad (\text{mod } 2).$$

Thus

$$[B_d : B_t] \begin{bmatrix} I_\mu \\ C_t^T \end{bmatrix} = 0,$$

and

$$B_d + B_t C_t^T = 0.$$

Also

$$B_t^{-1} \left[B_d + B_t C_t^T \right] = 0,$$

because B_t is nonsingular.

So

$$B_t^{-1} B_d = -C_t^T = C_t^T \quad (\text{mod } 2) \qquad (6.3)$$

By Corollary 6.1 we have

$$C_f K_f^T = 0 \quad (\text{mod } 2).$$

Thus

$$[K_d : I_{n-1}] \left[\frac{I_\mu}{C_t^T} \right] = 0,$$

and

$$K_d + C_t^T = 0.$$

So

$$K_d = -C_t^T = C_t^T \quad (\text{mod } 2).$$

Thus by (6.3),

$$K_d = B_t^{-1} B_d.$$

Let us use the graphs and matrices of Figures 6.2 and 6.3 to illustrate the above ideas. Let the spanning tree be $T = \{e_1, e_4, e_5\}$, and suppose that the fourth row of B in Figure 6.2(a) is deleted to create:

$$
B_r = [B_d : B_t] =
\begin{array}{ccccc}
e_2 & e_3 & e_1 & e_4 & e_5
\end{array}
\left[
\begin{array}{ccccc}
0 & 0 & : & 1 & 1 & 1 \\
1 & 0 & : & 1 & 0 & 0 \\
1 & 1 & : & 0 & 0 & 1
\end{array}
\right],
$$

$$
C_f = [I_2 : C_t] =
\begin{array}{ccccc}
e_2 & e_3 & e_1 & e_4 & e_5
\end{array}
\left[
\begin{array}{ccccc}
1 & 0 & : & 1 & 0 & 1 \\
0 & 1 & : & 0 & 1 & 1
\end{array}
\right], \quad \text{and}
$$

$$
K_f = [K_d : I_3] =
\begin{array}{ccccc}
e_2 & e_3 & e_1 & e_4 & e_5
\end{array}
\left[
\begin{array}{ccccc}
1 & 0 & : & 1 & 0 & 0 \\
0 & 1 & : & 0 & 1 & 0 \\
1 & 1 & : & 0 & 0 & 1
\end{array}
\right].
$$

Then

$$
C_t^T =
\begin{bmatrix}
1 & 0 \\
0 & 1 \\
1 & 1
\end{bmatrix}
= K_d.
$$

Also

$$
B_t^{-1} B_d =
\begin{bmatrix}
1 & 0 \\
0 & 1 \\
1 & 1
\end{bmatrix}
= C_t^T.
$$

From this we discover, for a spanning tree T, and the submatrix B_t of B_r corresponding to T that,

    (i) given B_r , we can construct C_f,
    (ii) given B_r , we can construct K_f ,
    (iii) given C_f , we can construct K_f , and
    (iv) given K_f , we can construct C_f .

A similar development can be made for the corresponding digraph matrices. We shall carry this out in Section 7.4.

6.5 The Path Matrix

The following matrix is useful in operations research (see Chapter 12) when one often wishes to determine paths in networks.

Let $G = (V, E)$ be a connected graph with its edges labelled.

Definition

For each pair of distinct vertices $u, v \in V$, the *path matrix* $P_{uv} = [p_{ij}]$, of a graph $G = (V, E)$, has a row for each path between u, v, and a column for each edge in E, in which $p_{ij} = 1$, if the j-th edge is a member of the i-th path between u and v in G, and $p_{ij} = 0$, otherwise.

To illustrate this, we turn once more to the graph in Figure 6.2(a). Let $u = v_1$ and $v = v_3$. There are three paths between v_1 and v_3, namely:

$$p_1 = \langle e_1, e_2 \rangle,$$
$$p_2 = \langle e_5 \rangle, \qquad \text{and}$$
$$p_3 = \langle e_4, e_3 \rangle.$$

This leads to:

$$P_{13} = \begin{array}{c} \begin{array}{ccccc} e_1 & e_2 & e_3 & e_4 & e_5 \end{array} \\ \begin{bmatrix} 1 & 1 & 0 & 0 & 0 \\ 0 & 0 & 0 & 0 & 1 \\ 0 & 0 & 1 & 1 & 0 \end{bmatrix} \end{array}. \tag{6.4}$$

We can make the following observations about P_{uv}.

P(i)   The sum of each row equals the number of edges in the path it represents.

P(ii)   A row of all unit entries represents a walk which is unicursal in the sense that it is an open Euler trail.

P(iii)   Each row must contain at least one unit entry.

P(iv)   A column with all unit entries represents an edge that belongs to every path between u and v.

P(v)   A column of all zeros represents an edge that does not lie on any path between u and v.

P(vi)   The sum of any two rows (mod 2) corresponds to an edge-disjoint union of cycles in G, which may possibly be a single cycle.

P(vii) $BP_{uv}^T = Q$ (mod 2), where Q has all zero entries except for exactly two rows which comprise solely unit entries.

In order to illustrate observation P(vii), we shall choose $u = v_1$ and $v = v_3$, giving rise to P_{uv}, as defined in Equation (6.4). From Figure 6.2(a),

$$BP_{13}^T = \begin{bmatrix} 1 & 0 & 0 & 1 & 1 \\ 1 & 1 & 0 & 0 & 0 \\ 0 & 1 & 1 & 0 & 1 \\ 0 & 0 & 1 & 1 & 0 \end{bmatrix} \begin{bmatrix} 1 & 0 & 0 \\ 1 & 0 & 0 \\ 0 & 0 & 1 \\ 0 & 0 & 1 \\ 0 & 1 & 0 \end{bmatrix}$$

$$= \begin{bmatrix} 1 & 1 & 1 \\ 0 & 0 & 0 \\ 1 & 1 & 1 \\ 0 & 0 & 0 \end{bmatrix} = Q.$$

6.6 Summary

It has been seen in this chapter that a number of matrices can be defined for graphs and digraphs. These matrices can be used to specify a graph in an exact manner. This is especially convenient in the storage of information about graphs on a mechanical device such as a computer. The adjacency and the incidence matrices specify a graph or a digraph to within isomorphism. However certain matrices can also be manipulated to analyse efficiently the properties of a given graph. This will be useful in certain sections of Part II of this book, especially in Chapter 13 on electrical engineering. In Chapter 7 we shall introduce the remaining matrices for digraphs which are analogous to those introduced for graphs in the present chapter.

6.7 Exercises

6.1 Prove that any cut-set and any cycle of a graph have an even number of edges in common.

6.2 Prove the unproved theorems stated in this chapter.

6.3 Prove that in a connected graph G, the complement of any cut-set in G does not contain a spanning tree.

6.4 Prove that in a connected graph G, the set of chords with respect to any spanning tree of G does not contain a cut-set.

6.5 Express the relationship of duality between two planar graphs in terms of appropriate matrices.

6.6 Characterize self-dual graphs in terms of their cycle and cut-set matrices.

6.7 In a connected graph G, let Q be a minimal set of edges with an even number (possibly zero) in common with every cut-set of G. Prove that Q is a cycle.

6.8 Prove the observations made throughout this chapter.

6.9 Prove that an Eulerian graph cannot have a cut-set with an odd number of edges.

6.10 Prove that the cyclomatic number of a graph does not change when either:

(i) a new vertex is inserted in the middle of one of its edges, or

(ii) a vertex of degree two is removed and its two incident edges are merged into a single edge.

7 Digraphs

By indirections find directions out.

Hamlet II, i — William Shakespeare

In Section 1.2 we defined the concept of a digraph and discussed some of its elementary properties. In Sections 6.1 and 6.2 we introduced the first two of the matrices which can be used to specify and analyse a digraph, namely the adjacency and incidence matrices. Because of their great applicability, we now explore the properties of digraphs in their own right, emphasizing those properties which set them apart from their analogous graph theoretic counterparts. We explore digraph connectivity, traversability, matrices, directed trees, the principle of directional duality, and tournaments.

7.1 Connectivity

Recall the definition of a digraph and related concepts from Section 1.2.

Suppose that $D = (V, A)$, is a digraph. Then a *walk* in D is an alternating sequence of vertices and arcs of $D : \langle v_0, a_1, v_1, \ldots, a_n, v_n \rangle$, in which each arc a_i is $v_{i-1}v_i$. The walk is termed *closed* if $v_0 = v_n$, and *spanning* if $\{v_0, v_1, \ldots, v_n\} = V$. A walk is termed a *trail* if all of its arcs are distinct, a *path* if all of its vertices are distinct, and a *cycle* if it contains at least two vertices and all its vertices are distinct except for the fact that $v_0 = v_n$. A digraph is said to be *cyclic* if it contains a cycle and *acyclic* otherwise. If there exists a path in D from one of its vertices u, to another v, then v is said to be *reachable from u*.

Definitions

If arc uv is a member of a digraph, its *converse* is defined to be arc vu.

The *converse digraph* D^c, of a digraph D, has the same vertex set as D, and an arc is in D^c if and only if its converse is in D.

A *semiwalk* is an alternating sequence: $\langle v_0, a_1, v_1, \ldots, a_n, v_n \rangle$ of vertices and arcs where each arc a_i is either: $v_{i-1}v_i$ or its converse. A semiwalk is termed a *semitrail* if all of its arcs are distinct, a *semipath* if all of its vertices are distinct, and a *semicycle* if it contains at least three vertices and all of its vertices are distinct except for the fact that $v_0 = v_n$.

We now discuss various types of connectivity in a digraph.

Definitions

A digraph is said to be *strongly connected* or *strong*, if every two of its distinct vertices, u and v say, are such that u is reachable from v and v is reachable from u. D is *unilaterally connected* or *unilateral,* if either u is reachable from v or v is reachable from u, and is *weakly connected, or weak,* if u and v are joined by a semipath.

Figure 7.1 shows: (a) a strong digraph, (b) a unilateral digraph, and (c) a weak digraph.

While every strongly connected digraph is unilateral and every unilateral digraph is weak, there are weak digraphs which are not unilateral, and unilateral digraphs which are not strong.

A digraph is said to be *disconnected* if it is not weak.

The following theorems are useful in analysing the connectivity of digraphs.

Theorem 7.1

A digraph is strong if and only if it has a spanning closed walk.

Proof

(Necessity)

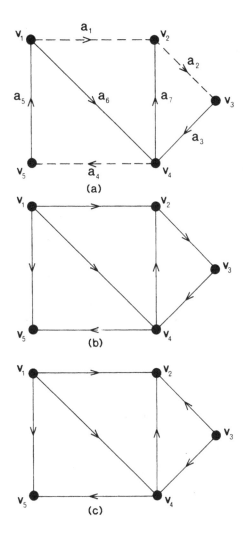

Figure 7.1 Strong, unilateral and weak digraphs

If $D = (V, A)$ is a strong digraph with $V = \{v_1, v_2, \ldots, v_n\}$, there is a walk from each vertex in V to each other vertex in V. Thus there exist in D, walks: $W_1, W_2, \ldots, W_{n-1}$ such that the first vertex of W_i is v_i and the last vertex of W_i is v_{i+1}, for $i = 1, 2, \ldots, n - 1$. There also exists a walk, say W_n, with first vertex v_n and last vertex v_1. Then the walk obtained by traversing the walks: $W_1, W_2, \ldots, W_n$ in succession, is a spanning closed walk of D.

(Sufficiency)
Let u and v be two distinct vertices of V. If v follows u in any spanning closed walk, say W, of D then there exists a sequence of the arcs of W constituting a walk from u to v. If u follows v in W, then there is a walk from u to the last vertex of W and a walk from that vertex to v. A walk from u to v can be constructed by traversing these two walks in succession. ∎

Theorem 7.2

A digraph D is unilateral if and only if it has a spanning walk.

Proof

(Necessity)
Suppose D is a unilateral digraph. Let W be a walk in D containing the maximum number of vertices. Suppose that W begins at vertex v_1 of D and ends at vertex v_2 of D. If W is a spanning walk the proof is completed. Assume that W is not a spanning walk. Then there exists a vertex, u say, of D that is not in W. Also there cannot exist in D a walk from u to v_1 or a walk from v_2 to u. Because D is unilateral and does not possess a walk from u to v_1, D must possess a walk from v_1 to u.

Let t (which cannot be v_2) be the last vertex of W from which a walk from t to u exists in D. Let U be a walk from t to u in D. Let s be the vertex in D which is the immediate successor of the last appearance of t in W. D does not possess a walk from s to u, however because D is unilateral, there is a walk say, Y, from u to s in D. Let us traverse W from v_1 to the last appearance of t, then traverse u, then traverse Y to vertex s and then traverse W to v_2. This represents a walk from v_1 to v_2 which has more distinct vertices than W, which is a contradiction. Hence W is a spanning walk of D.

(Sufficiency)
Obvious. ∎

If $D = (V, A)$, is a unilateral digraph then $od(u) = 0$ for at most one $u \in V$ and $id(v) = 0$ for at most one $v \in V$, $u \neq v$.

Theorem 7.3

A digraph is weak if and only if it has a spanning semiwalk.

Proof

(Necessity)
Let $D = (V, A)$, be a weak digraph with $V = \{v_1, v_2, \ldots, v_n\}$. Because D is weak there is a semiwalk, say W_i, from v_i to v_{i+1} in D for $i = 1, 2, \ldots, n-1$. The semiwalk obtained by traversing the semiwalks: $W_1, W_2, \ldots, W_{n-1}$, in succession is a spanning semiwalk of D.

(Sufficiency)
Let D be a digraph containing a spanning semiwalk say W.

Let v_1 and v_2 be any two distinct vertices of D. Naturally v_1 and v_2 belong to W as it is spanning. The part of W which begins at any appearance of $v_1(v_2)$ and ends at any appearance of $v_2(v_1)$ represents a semiwalk from v_1 to v_2 (from v_2 to v_1) in D. Thus there is either a semiwalk from v_1 to v_2 or from v_2 to v_1 in D. Thus D is weak. ■

Definition

A digraph $D' = (V', A')$, is termed a *subdigraph* of a digraph $D = (V, A)$, if $V' \subseteq V$ and $A' \subseteq A$ and D' is a digraph.

Just as there are three concepts of connectivity in the theory of digraphs there are also three kinds of components.

Definitions

A *strong component* in a digraph D, is a maximal strong subdigraph of D.

A *unilateral component* in a digraph D, is a maximal unilateral subdigraph of D.

A *weak component* in a digraph D, is a maximal weak subdigraph of D.

These concepts are now illustrated. The digraph in Figure 7.1(b) has a strong component induced by the vertex set $\{v_2, v_3, v_4\}$. The digraph in Figure 7.1(c) has a unilateral component induced by the vertex set $\{v_1, v_4, v_2\}$, and a weak component which is the digraph itself.

Every vertex and every arc of a digraph D, belongs to exactly one weak component of D and to at least one unilateral component of D. Also, every vertex belongs to exactly one strong component of D. Every arc a, belongs to exactly one strong component of D if a belongs to a cycle of D, and to no strong component of D if a does not belong to a cycle of D.

Definition

Let D be a digraph with strong components: $S_1, S_2, \ldots, S_p$. The *condensation* $D^* = (V^*, A^*)$, of D is the digraph with $V^* = \{S_1, S_2, \ldots, S_p\}$ and where $S_i S_j$ is an arc of A^* if and only if there exists an arc uv in D for $u \in S_i$ and $v \in S_j$.

By way of illustration, the condensation of the digraph in Figure 7.1(a) is just a single vertex. The condensation of the digraph in Figure 7.1(b) comprises $V^* = \{S_1, S_2, S_3\}$, and $A^* = \{(S_1, S_2), (S_1, S_3), (S_2, S_3)\}$, where $S_1 = \{v_1\}$, $S_2 = \{v_2, v_3, v_4\}$ and $S_3 = \{v_5\}$. The condensation of the digraph in Figure 7.1(c) is the digraph itself.

These illustrations are examples of some general observations. Let D^* be the condensation of a digraph D.

(i) D^* is acyclic.
(ii) If D^* is strong then D^* comprises a single vertex (and no arcs).
(iii) D^* comprises a unique spanning path and is unilateral if and only if D is unilateral.

Definition

A *cut-set* in a digraph $D = (V, A)$, is a set of arcs of A, which constitute a cut-set in the multigraph $G = (V, E)$, obtained from D by removing the orientation from each arc of A.

We turn now to the concept of digraph traversability.

7.2 Traversability

Definition

A digraph D, is said to be *Eulerian* if it contains a closed trail which traverses every arc of D exactly once. Such a trail is termed an *Euler* trail. D is said to be *unicursal* if it contains an open Euler trail. It is natural to ask which digraphs are Eulerian and which are unicursal. We now state and prove a theorem which, with its corollary, is analogous to Theorems 4.1 and 4.2 for multigraphs.

Theorem 7.4

A digraph $D = (V, A)$, is Eulerian if and only if D is connected and for each of its vertices v, $id(v) = od(v)$.

Proof

(Necessity).
Suppose that D is an Eulerian digraph. It therefore contains an Eulerian trail, say T. In traversing T, every time a vertex v, is encountered we pass along an arc incident towards v and then an arc incident away from v. This is true for all the vertices of T, including the initial vertex of T, say v, because we began T by traversing an arc incident away from v, and ended T by traversing an arc incident towards v.

(Sufficiency).
Assume that all of the vertices of D have the same number of arcs incident towards them as incident away from them. Choose an arbitrary vertex, say v, in D. We identify a trail starting at v which traverses the arcs of D at most once each. We shall continue traversing the arcs of D until it is impossible to traverse further. Because every vertex has the same number of arcs incident towards it as away from it, we can leave any vertex that we enter via the trail and the traversal will not stop until v is reached. However, since v also has the assumed property, the traversal will eventually reach v and there it will be impossible to traverse further. Let the trail traversed so far be denoted by T. If it includes all of A, then the result follows. If not, we now remove from D all the arcs of T and consider the remainder of A. By assumption, each vertex in the remaining digraph, say D', is such that the number of arcs directed towards it equals the number of arcs directed away from it. Furthermore, T and D', must have a vertex, say u, in common because D is connected. Beginning at u, we repeat the process of tracing a trail in D'. If this trail does not encounter all of the arcs of D', the process can be repeated until a closed trail that traverses each of the arcs of D exactly once is produced. Thus D is Eulerian. ■

Corollary 7.1

Let $D = (V, A)$, be a weak digraph. D is unicursal if and only if D contains vertices u and v such that: $od(u) = id(u) + 1$, $id(v) = od(v) + 1$, and $od(w) = id(w)$, for all $w \in V$ where $w \neq u, v$. In this case D has an open Euler trail which begins at u and ends at v.

We now make a brief excursion into Hamiltonian digraphs.

Definition

A digraph D is termed *Hamiltonian* if it has a cycle containing all of the vertices of D.

There are a number of results that are analogous to those proved for Hamiltonian graphs in Section 4.3. For instance, let D be a strong digraph with

n vertices $(n \geq 3)$ such that for every pair u and v, of distinct non-adjacent vertices of D, $d(u) + d(v) \geq 2n - 1$. It has been shown by Meyniel (1973) that D is Hamiltonian.

7.3 Directed Trees

Recall that a tree, as defined in Section 3.1, is a connected acyclic graph. A directed tree can be defined analogously.

Definition

A *directed tree* is a weak digraph that does not contain a semicycle.

There is one particular type of directed tree that is of importance in network analysis, computer science, enumeration, and other fields of applied graph theory. It is called an *arborescence*.

Definition

A directed tree is said to be an *arborescence* if it contains exactly one vertex, called the *root*, with no arcs directed towards it, and if all the arcs on any semipath are directed away from the root.

One can make a number of observations about any arborescence D:

(i) Every vertex in D, other than the root, has exactly one arc directed towards it.

(ii) There is a path from the root of D to every other vertex in D.

(iii) The root r, of D has the property that every other vertex in D is reachable from r, and r is not reachable from any other vertex of D.

7.4 More Digraph Matrices

In Sections 6.1 and 6.2 we introduced the adjacency and incidence matrices of a digraph. We now discuss digraph matrices corresponding to the matrices for graphs introduced in Sections 6.3, 6.4, and 6.5.

Let $D = (V, A)$, be a digraph whose arcs have been labelled and whose semicycles have been both labelled and given an arbitrary orientation.

Definition

The *semicycle matrix* $C = (c_{ij})$, of a digraph D is a matrix in which $c_{ij} = 1$ if semicycle Z_i of D is such that arc a_j is directed in the same way as the orientation of Z_i, $c_{ij} = -1$ if Z_i contains arc a_j directed in the opposite way to the orientation of Z_i, and $c_{ij} = 0$ otherwise.

As with graphs the arcs of a connected digraph D, not part of a given spanning directed tree T, of D are called *chords*. When a chord is added to T it creates a *fundamental semicycle* (which may be a cycle).

We now illustrate these ideas with the digraph in Figure 7.1(a). Here T is specified by the arcs in solid lines, with branches: a_3, a_5, a_6, and a_7. The chords with respect to T are shown as the dotted lines: a_1, a_2, and a_4. The fundamental semicycles with respect to T are:

Chord	Semicycle	
a_1	$\langle a_1, a_7, a_6 \rangle$,	
a_2	$\langle a_2, a_3, a_7 \rangle$,	(a cycle),
a_4	$\langle a_4, a_5, a_6 \rangle$,	(a cycle).

The rows of the semicycle matrix C of a digraph D, containing a specified spanning tree T, which correspond to its fundamental semicycles with respect to T, constitute a submatrix of C, called the *fundamental semicycle matrix C_f* (with respect to T).

If D has n vertices and e arcs then there are $\mu = e - n + 1$ fundamental semicycles with respect to any specified spanning tree. As with graphs, it is possible to generate all the semicycles of D from linear combinations of the fundamental semicycles. This is usually carried out by row manipulation of C_f, where we use ordinary (rather than modulo 2) arithmetic.

A natural orientation for the fundamental semicycles is provided by the direction of the chords which give rise to them, with respect to a given spanning tree. We now illustrate this for the graph in Figure 7.1(a), with T defined as before. Then

$$C_f = \begin{array}{c} Z_1 \\ Z_2 \\ Z_3 \end{array} \begin{bmatrix} 1 & 0 & 0 & 0 & 0 & -1 & -1 \\ 0 & 0 & 0 & 1 & 1 & 1 & 0 \\ 0 & 1 & 1 & 0 & 0 & 0 & 1 \end{bmatrix} .$$

The cycle matrix C, for this digraph can be generated by taking linear combinations of C_f. Note that, although it is termed the cycle matrix, some of its rows represent semicycles which are not cycles, as indicated by (-1) entries.

$$
C = \begin{array}{c} Z_1 \\ Z_2 \\ Z_3 \\ Z_4 \\ Z_5 \\ Z_6 \end{array}
\begin{bmatrix}
1 & 0 & 0 & 0 & 0 & -1 & -1 \\
0 & 0 & 0 & 1 & 1 & 1 & 0 \\
0 & 1 & 1 & 0 & 0 & 0 & 1 \\
1 & 1 & 1 & 0 & 0 & -1 & 0 \\
1 & 0 & 0 & 1 & 1 & 0 & -1 \\
1 & 1 & 1 & 1 & 1 & 0 & 0
\end{bmatrix}
\begin{array}{l} \\ \\ \\ (= Z_1 + Z_3) \\ (= Z_1 + Z_2) \\ (= Z_1 + Z_2 + Z_3). \end{array}
$$

We now make some observations about C for an arbitrary digraph D.

(i) An arc in D that does not belong to any semicycle is represented by a column of zeros.

(ii) The number of nonzero entries in a row r, equals the number of arcs in the semicycle that r represents.

It is also possible to prove a result which is analogous to Theorem 6.1.

Theorem 7.5

Let D be a digraph with incidence matrix B, and semicycle matrix C, in which the columns are arranged according to the same order of arcs. Then

$$BC^T = 0,$$

and

$$CB^T = 0,$$

where the matrix arithmetic is carried out in the field of real numbers.

Proof

Consider the p-th row of B and the q-th row of C. The q-th semicycle, say c_q, either: (a) does not, or (b) does possess an arc incident with vertex, say v_p, represented by the p-th row of B. If (a), the product of the two rows is zero. If (b), there are exactly two arcs, say a_i, and a_j, of the q-th semicycle incident with v_p. There are four possibilities:

(i) a_i and a_j are both incident towards v_p,

(ii) a_i and a_j are both incident away from v_p,

(iii) The directions of both a_i and of a_j are compatible with the orientation of c_q, and

(iv) The directions of both a_i and of a_j are incompatible with the orientation of c_q.

It is straightforward to check that, in all four cases, the product of the p-th row of B and the q-th row of C is zero. ∎

Definition

The *rank* of a digraph is defined to be the rank of its incidence matrix.

We can make some further observations about the relationship between B and C for a digraph with n vertices and e arcs, where the rank of a matrix M, is denoted by $r(M)$:

(i) $r(B) + r(C) = e$,

(ii) $r(B) = n - 1$ for any weak digraph, and

(iii) $r(C) = e - n + 1$. (From observations (i) and (ii).)

Let us now return to C_f for the digraph in Figure 7.1(a) and permute its columns so as to create a matrix of the form: $[I_\mu : C_t]$.

$$
C_f = \begin{array}{c} \begin{array}{ccccccc} a_1 & a_4 & a_2 & & a_5 & a_6 & a_7 & a_3 \end{array} \\ \left[\begin{array}{ccccccc} 1 & 0 & 0 & : & 0 & -1 & -1 & 0 \\ 0 & 1 & 0 & : & 1 & 1 & 0 & 0 \\ 0 & 0 & 1 & : & 0 & 0 & 1 & 1 \end{array} \right] \end{array}.
$$

As with the cycle matrix, we can define the *cut-set matrix* $K = (k_{ij})$, for any weak digraph $D = (V, A)$, in which the rows correspond to the cut-sets of D and the columns to the arcs of D. Each cut-set must be given an (arbitrary) orientation as follows. Let k_i be the i-th cut-set of D. Suppose that k_i partitions V into nonempty vertex sets V_i' and V_i''. The *orientation* can be defined to be either: from V_i' to V_i'' or from V_i'' to V_i'. Suppose that the orientation is chosen to be from V_i' to V_i''. Then the *orientation* of an arc a_j of cut-set k_i, is said to be the *same* as that of k_i if a_j is of the form $v_a v_b$ where $v_a \in V_i'$ and $v_b \in V_i''$ and the *opposite*, otherwise. Then $k_{ij} = 1$, if arc a_j of cut-set k_i has the same orientation as k_i, $k_{ij} = -1$, if arc a_j of cut-set k_i has the opposite orientation to k_i, and $k_{ij} = 0$, otherwise.

We can make some observations about K, the cut-set matrix of a weak digraph D, with n vertices and e arcs.

(i) A permutation of the rows or columns corresponds to a relabelling of the cut-sets and arcs of D respectively.

(ii) $r(K) \geq r(B)$,

(iii) $r(K) \geq n - 1$, (By observation (ii).)

(iv) If the arcs of D are arranged in the same column order in C and K, then

$$
CK^T = 0,
$$

and

$$
KC^T = 0,
$$

where the matrix arithmetic is carried out in the field of real numbers.

(v) $r(C) + r(K) \leq e$,

(vi) Because D is weak, $r(C) = e - n + 1$, and $r(K) \leq n - 1$. Hence

(vii) $r(K) = n - 1$. (By observations (iii) and (iv).)

As with the semicycle matrix, we shall find it convenient to define a *fundamental cut-set matrix* (which has linearly independent rows) of a digraph D.

As with graphs, the removal of an arc, say $a = v_p v_q$, (also called a *branch*) of a spanning directed tree of D, partitions the vertices of a digraph D into two disjoint sets, say V_1 and V_2.

The cut-set created by the removal of a is said to be either:

(i) *directed away from V_1 and towards V_2* if $v_p \in V_1$ and $v_q \in V_2$, or

(ii) *directed away from V_1 and towards V_2* if $v_p \in V_2$ and $v_q \in V_1$.

Such a cut set is termed a *fundamental cut-set*. Of course not all the chords in k_i necessarily have the same orientation as $v_q v_p$. If $v_q v_p$ is directed away from a vertex in V_1, there may exist a chord in k_i which is directed towards a vertex in V_1. The orientation of a cut-set on the basis of the direction of the branch giving rise to it constitutes a natural way of orienting cut-sets. If all the chords of k_i are oriented as $v_q v_p$ is, then k_i is said to be *directed*.

Returning to the digraph in Figure 7.1(a), with T defined as before, the fundamental cut-sets with respect to T are:

branch	cut-set
a_3	$\{a_3, a_2\}$
a_5	$\{a_5, a_4\}$
a_6	$\{a_6, a_1, a_4\}$
a_7	$\{a_7, a_1, a_2\}$.

A fundamental cut-set is created from K, the cut-set matrix of a weak digraph with given directed spanning tree T, by deleting from K, all rows which do not correspond to fundamental cut-sets with respect to T. Therefore K_f is an $(n - 1) \times e$ submatrix of K such that each row represents a unique fundamental cut-set with respect to T.

As with the fundamental semicycle matrix, the rows of any fundamental cut-set matrix K_f, can be permuted to create a matrix of the form: $K_f = [K_c : I_{n-1}]$. Here K_c is an $(n - 1) \times (e - n + 1)$ matrix whose columns correspond to the chords of T and I_{n-1} is the identity matrix of order $n - 1$ whose columns correspond to the branches of T. Let D be a weak digraph with given directed spanning tree T. We now deduce relationships among C_f, K_f, and B_r. (Here B_r is the reduced incidence matrix in which an arbitrary row has been removed in order to make its rows linearly independent.) We have seen that

$$C_f = [I_\mu : C_t], \tag{7.1}$$

and

$$K_f = [K_c : I_{n-1}], \tag{7.2}$$

where t corresponds to the branches of T and c to the chords of T. We assume that the arcs are assembled in the same order in (7.1), (7.2), and in B_r. We begin by partitioning B_r into:

$$B_r = [B_c : B_t],$$

where B_c is an $(n-1) \times (e-n+1)$ submatrix whose columns correspond to the chords of T and B_t is an $(n-1) \times (n-1)$ submatrix whose columns correspond to the branches of T. Because

$$BC^T = 0,$$

we can deduce, by analogous reasoning to that used following Theorem 6.7, that

$$K_c + C_t^T = 0.$$

Therefore

$$K_c = -C_t^T, \tag{7.3}$$

and thus

$$K_c = B_t^{-1} B_c. \tag{7.4}$$

In order to illustrate these relationships we consider once again the digraph in Figure 7.1(a), with the spanning tree T, defined as before.

$$B_r = [B_c : B_t] = \begin{array}{cccccccc} a_1 & a_4 & a_2 & & a_5 & a_6 & a_7 & a_3 \\ \begin{bmatrix} 1 & 0 & 0 & : & -1 & 1 & 0 & 0 \\ -1 & 0 & 1 & : & 0 & 0 & -1 & 0 \\ 0 & 0 & -1 & : & 0 & 0 & 0 & 1 \\ 0 & 1 & 0 & : & 0 & -1 & 1 & -1 \end{bmatrix} \end{array}.$$

$$C_f = [I_3 : C_t] = \begin{array}{c} \begin{array}{ccccccc} a_1 & a_4 & a_2 & a_5 & a_6 & a_7 & a_3 \end{array} \\ \left[\begin{array}{ccccccc} 1 & 0 & 0 & : & 0 & -1 & -1 & 0 \\ 0 & 1 & 0 & : & 1 & 1 & 0 & 0 \\ 0 & 0 & 1 & : & 0 & 0 & 1 & 1 \end{array} \right] \end{array}.$$

$$K_f = [K_c : I_4] = \begin{array}{c} \begin{array}{ccccccc} a_1 & a_4 & a_2 & a_5 & a_6 & a_7 & a_3 \end{array} \\ \left[\begin{array}{ccccccc} 0 & -1 & 0 & : & 1 & 0 & 0 & 0 \\ 1 & -1 & 0 & : & 0 & 1 & 0 & 0 \\ 1 & 0 & -1 & : & 0 & 0 & 1 & 0 \\ 0 & 0 & -1 & : & 0 & 0 & 0 & 1 \end{array} \right] \end{array}.$$

Note that the last row of B, corresponding to vertex v_5, has been removed to create B_f. We shall form linear combinations of the rows of K_f to create all 10 rows of K, representing all of the cut-sets of the digraph in Figure 7.1(a).

$$K = \begin{array}{c} \begin{array}{ccccccc} a_1 & a_4 & a_2 & a_5 & a_6 & a_7 & a_3 \end{array} \\ \left[\begin{array}{ccccccc} 0 & -1 & 0 & 1 & 0 & 0 & 0 \\ 1 & -1 & 0 & 0 & 1 & 0 & 0 \\ 1 & 0 & -1 & 0 & 0 & 1 & 0 \\ 0 & 0 & -1 & 0 & 0 & 0 & 1 \\ -1 & 0 & 0 & 1 & -1 & 0 & 0 \\ 0 & -1 & 0 & 0 & 1 & -1 & 1 \\ 0 & 0 & 1 & -1 & 1 & -1 & 0 \\ 0 & 0 & 0 & 1 & 1 & 1 & -1 \\ 0 & -1 & 1 & 0 & 1 & -1 & 0 \\ 1 & 0 & 0 & 0 & 0 & 1 & -1 \end{array} \right] \begin{array}{l} k_1 \\ k_2 \\ k_3 \\ k_4 \\ k_1 - k_2 \\ k_2 - k_3 + k_4 \\ k_2 - k_1 - k_3 \\ k_1 - k_2 + k_3 - k_4 \\ k_2 - k_3 \\ k_3 - k_4 \end{array} \end{array}.$$

Thus we can make similar observations to those made in Section 6.4.

1. Given B_r, we can construct C_f.

2. Given B_r, we can construct K_f.

3. Given C_r, we can construct K_f.

4. Given K_f, we can construct C_f.

We define a *semipath* matrix for a digraph analogously to that for the path matrix of a graph.

Definition

The *semipath matrix* $P(u, v) = (p_{ij})$, of a digraph $D = (V, A)$, where $u, v \in V$, is the matrix with each row representing a distinct semipath from u to v and the columns representing the arcs of D, in which $p_{ij} = 1$ if the i-th semipath contains the j-th arc, $p_{ij} = -1$ if the i-th semipath contains the converse of the j-th arc, and $p_{ij} = 0$ otherwise.

The matrix $P(v_3, v_5)$, for the digraph of Figure 7.1(a) is:

$$
P(v_3, v_5) = \begin{array}{c} \begin{array}{ccccccc} a_1 & a_2 & a_3 & a_4 & a_5 & a_6 & a_7 \end{array} \\ \left[\begin{array}{ccccccc} 1 & 1 & 0 & 0 & 1 & 0 & 0 \\ 1 & 0 & -1 & 0 & 1 & 0 & -1 \\ 0 & 1 & 0 & 0 & 1 & 1 & 1 \\ 0 & 0 & -1 & 0 & 1 & 1 & 0 \\ 1 & 1 & 0 & -1 & 0 & -1 & 0 \\ 1 & 1 & 0 & -1 & 0 & -1 & 0 \\ 0 & 1 & 0 & -1 & 0 & 0 & 1 \end{array} \right] \end{array}.
$$

We can make a number of observations about P.

1. If $P(u, v)$ contains a column of all zeros then the vertex that it represents does not belong to any of the semipaths between u and v.

2. If $P(u, v)$ contains a column of all unit entries then the vertex that it represents belongs to every semipath between u and v.

3. The number of nonzero entries in any row of $P(u, v)$ equals the number of arcs in the semipath represented by the row.

7.5 The Principle of Directional Duality

There is a very powerful principle in the theory of binary relations which has direct and useful application in the analysis of digraphs.

We now state a specialization of the principle of directional duality for digraphs:

The Principle of Directional Duality for Digraphs.

For many theorems on digraphs there is an analogous theorem obtained by replacing each concept in the theorem by its converse.

In order to illustrate this principle we first state,

Theorem 7.6

An acyclic digraph has at least one vertex with no arcs incident towards it.

Proof

Consider the first vertex of any maximal path in the digraph. This vertex cannot have any arcs adjacent towards it, otherwise the path would not be maximal or the digraph would contain a cycle. Hence the result follows. ■

Theorem 7.7

An acyclic graph has at least one vertex with no arcs incident away from it.

Proof

The proof follows easily by applying the principle of directional duality to the proof of Theorem 7.6. ■

7.6 Tournaments (Based on Chapter 4, Robinson and Foulds (1980).)

In popular speech, a tournament usually means some kind of sporting competition. We use the word in a way which corresponds to what are frequently called *round-robin tournaments*, where each pair of contestants (individual or team) play each other exactly once. We assume that exactly one of the pair wins. Such sporting tournaments (and the ranking of objects in the social sciences by comparing them two at a time) can be modelled by the theory of digraphs as follows. Each contestant is represented by a unique vertex of a digraph D. If the competitor represented by vertex v_i, beats the competitor represented by vertex v_j, then D will possess arc $v_i v_j$. This gives rise to the following definition.

Definition

A *tournament* is an oriented complete graph.

Figure 7.2 shows all the tournaments on 1, 2, 3, or 4 vertices.

Theorem 7.8

For any vertex v, in a tournament with n vertices,
$$id(v) + od(v) = n - 1.$$

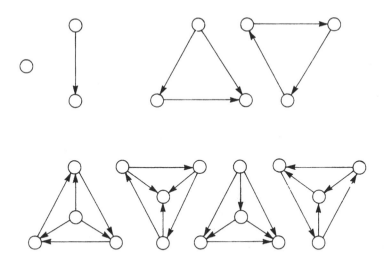

Figure 7.2 All tournaments with at most four vertices

Proof

There is exactly one arc between u and each of the $n - 1$ other vertices.
∎

Corollary 7.2

Any tournament with n vertices has $\frac{1}{2}n(n - 1)$ arcs.

If u and v are vertices in any tournament T then either (u, v) or (v, u) is an arc of T. Thus either $< u, v >$ or $< v, u >$ is a walk in T and hence either v is reachable from u or u is reachable from v. Therefore T is unilateral.

Because every tournament is unilateral, the results of Section 7.1 may be applied to tournaments. We shall also find that the extra properties of tournaments enable us to strengthen these results considerably. The results of Section 7.1 show that the strong components of a tournament $D = (V, A)$, may be placed in order: $C_1, C_2, \ldots, C_k$ in such a way that if $u \in C_i$ and $v \in C_j$, and if $i > j$, then $(u, v) \in A$. Hence if we know the components and their order, we know the results of each comparison between objects in different components. The special nature of tournaments leads to a theorem on the possible size of a component.

Theorem 7.9

No component in a tournament consists of exactly two vertices.

Proof

In any digraph, $D = (V, A)$, with $u, v \in V$, $\{u, v\}$ is a component if and only if there is a walk from u to v and a walk from v to u, neither passing through any other vertex. This requires that both (u, v) and (v, u) be arcs. But these two arcs cannot both appear in a tournament. ∎

The component structure of a tournament is therefore, to a considerable extent, described by the cardinalities of its components and the sequence in which they occur. Each tournament has a *component size sequence*, consisting of the cardinalities of the components in the order in which they occur in the spanning path of the condensation of the tournament. The sum of these numbers must be the number of vertices in the tournament and the number 2 cannot occur. These are the only restrictions. The tournaments in Figure 7.2 have component cardinality sequences:

$$1; \quad 1, 1; \quad 1, 1, 1; \quad 3; \quad 1, 3; \quad 3, 1; \quad 1, 1, 1, 1; \text{ and } 4,$$

respectively.

The nature of tournaments also implies that subdigraphs of a tournament share that property, as stated in the following theorem.

Theorem 7.10

Consider a subdigraph S of any tournament and the underlying graph G, say, obtained from S by removing the orientations of the arcs of S. If G is complete then S is a tournament.

Proof

From the definition of a tournament. ∎

Definition

A subdigraph of a tournament is called a *subtournament*.

It is an immediate consequence of Theorem 7.10 that every component of a tournament is a subtournament. It is also strong. Hence we now examine strong tournaments.

7.6.1 Strong Tournaments

The first, fourth, and last tournaments in Figure 7.2 are strong. Such tournaments exist for any number of vertices except two.

Theorem 7.11

If T is a strong tournament with n vertices, and $3 \leq k \leq n$, then there is a cycle in T which passes through exactly k vertices.

Proof

Let u be a vertex of T. Then the vertices of T can be divided into three sets: $\{u\}$, W, and X, where W consists of those vertices w, such that (w, u) is an arc, and X those vertices x such that (u, x) is an arc. As T is strong (and has at least three vertices by assumption), neither W nor X is empty. If there is no arc from any vertex in X to any vertex in W, then no non-trivial closed walk can pass through u, for it would be of the form $\langle u, x, \ldots, w, u \rangle$, and there is no walk from X to W. Hence T must have vertices $w \in W$ and $x \in X$, such that (x, w) is an arc of T. Then $\langle u, x, w, u \rangle$ is a 3-cycle in T. Now suppose that we have a cycle $\langle u_1, u_2, \ldots, u_k, u_1 \rangle$ in T, and that there is at least one vertex of T not in the set $U_k = \{u_1, u_2, \ldots, u_k\}$. First, there may be a vertex $v \in V \backslash U_k$, such that for some i, (u_i, v) and (v, u_{i+1}) are both arcs of T (where u_{k+1} is interpreted as u_1). Then $\langle u_1, u_2, \ldots, u_i, v, u_{i+1}, \ldots, u_k, u_1 \rangle$ is a cycle through $k + 1$ vertices.

On the other hand there may not be any such vertex v. In this case, each vertex not in U_k is such that either : all members of U_k are its successors, or all are its precursors. We then divide $V \backslash U_k$ into two sets: W_k and X_k. For $w \in W_k$, (w, u_i) is an arc for each $i \in \{1, 2, \ldots, k\}$, and for $x \in X_k$, (u_i, x) is an arc for each $i \in \{1, 2, \ldots, k\}$. We know that $W_k \cup X_k \neq \emptyset$, for otherwise $U_k = V$, so neither W_k nor X_k can be empty, for that would prevent T from being strong. Further, T can be strong only if there is a closed walk through members of X_k and W_k (and U_k).

Since no arc begins in U_k and ends in W_k or begins in X_k and ends in U_k, there must be an arc beginning in X_k and ending in W_k. Let (x, w) be such an arc. Then (leaving out u arbitrarily), $\langle u_1, x, w, u_3, \ldots, u_k \rangle$ is a cycle of length $k + 1$. (We do not have to leave out any vertex at all to make a cycle. Also we have a free choice as to which vertex to leave out. Since we are introducing two new vertices, unless we leave one out there might be gaps in the list of values of k for which cycles of length k exist). The process of increasing the number of vertices can be continued until $k = n$. Hence the result. ■

We may observe that the details of the proof of Theorem 7.11 require that, in constructing the spanning cycle from the cycle through $p-1$ vertices, there must be a vertex v, of the kind required, with (u_i, v) and (v, u_{i+1}) both arcs. To illustrate this theorem we consider the strong tournament shown in Figure 7.3. The proof of the theorem actually demonstrates more than the statement asked. It says that we can choose any vertex of the tournament and there will be a cycle of length k, for each k, $3 \le k \le n$, through that vertex. In Figure 7.3 we choose vertex 1 and find cycles of lengths: 3, 4, 5, 6, and 7 through vertex 1.

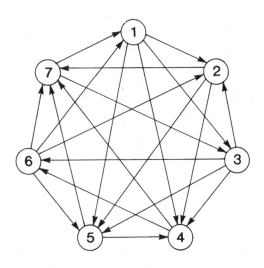

Figure 7.3   A strong tournament

To find our initial cycle of length 3 through vertex 1, let

$$W = \{6, 7\} \text{ and } X = \{2, 3, 4, 5\}.$$

We observe that $(2, 7)$ is an arc beginning in X and ending in W, so a cycle of length 3 is

$$\langle 1, 2, 7, 1 \rangle,$$

with $k = 3$, $u_1 = 1$, $u_2 = 2$, and $u_3 = 7$. We notice that $(1,3)$ and $(3,2)$ are both arcs, giving the cycle

$$\langle 1, 3, 2, 7, 1 \rangle.$$

At the next stage there is no $v \in \{4, 5, 6\}$ such that $(1, v)$ and $(v, 2)$ are both arcs, but $(3, 6)$, $(6, 2)$ are arcs, so

$$\langle 1, 3, 6, 2, 7, 1 \rangle$$

is the next cycle. Then $(3, 4)$ and $(4, 6)$ are arcs, and the next cycle is

$$\langle 1, 3, 4, 6, 2, 7, 1 \rangle.$$

Finally $(3, 5)$ and $(5, 4)$ are arcs so that

$$\langle 1, 3, 5, 4, 6, 2, 7, 1 \rangle$$

is the spanning cycle.

Theorem 7.11 allows us to strengthen Theorem 7.2 that a unilateral digraph has a spanning walk, for the special case of tournaments.

Theorem 7.12

Every tournament has a spanning path.

Proof

Each component of a tournament is a strong tournament and has a spanning cycle. If component C_i, has a single vertex, this cycle is $\langle x \rangle$, where $\{x\} = C_i$, otherwise it has the form $\langle x_i, \ldots, y_i, x_i \rangle$.

Moreover, (y_i, x_{i+1}) is an arc for each pair of consecutive components C_i, C_{i+1}. If we use y_i as an alternative name for x_i in components with only one vertex,

$$\langle x_1, \ldots, y_1, x_2, \ldots, y_2, \ldots, x_k, \ldots, y_k \rangle$$

is a spanning path of the tournament. ∎

Tournaments can be applied in the scheduling of round-robin sporting competitions in those sports in which each competing team has its own *home* ground. For fairness, we may want each team to play half its matches at home and half away. We now discuss how to attempt to arrange this. Because the number of matches each competitor plays is one less than the number, say n, of competitors, this will only be possible if the number of competitors is odd. One rule is to assign each competitor a unique integer between 1 and n. Then if i and j are two competitors, their match is played on i's home ground if the difference $|i - j|$ is odd and at j's home ground if $|i - j|$ is even. It is apparent that this gives each team half its matches at home and half away. When $p = 7$, the matches are (home competitor, away competitor): (1,2), (3,1), (1,4), (5,1), (1,6), (7,1), (2,3), (4,2), (2,5), (6,2), (2,7),(3,4), (5,3), (3,6), (7,3), (4,5), (6,4), (4,7), (5,6), (7,5), (6,7).

7.6.2. Acyclic Tournaments

At the other extreme from strong tournaments, a tournament may be acyclic, as are the first, second, third and fifth tournaments of Figure 7.2. Acyclic tournaments are important, since they correspond to consistent rankings, and are easily detected.

Definition

A *logical numbering* of a digraph $D = (V, A)$, with n vertices is a one-to-one correspondence $f : V \rightarrow \{1, 2, \ldots, n\}$, such that if $uv \in A$ then $f(u) < f(v)$.

Theorem 7.13

An acyclic tournament $T = (V, A)$, with n vertices has only one logical numbering and one spanning path. If f is the logical numbering of T and if $f(v) = i$ for some $v \in V$. Then $od(v) = n - i$ and $id(v) = i - 1$.

Proof

Every tournament has a spanning path by Theorem 7.12. Let

$$\langle v_1, v_2, \ldots, v_p \rangle$$

be a spanning path of an acyclic tournament. Then if we let v_i have number i, the numbering is consistent with the arcs of the spanning path, and since the digraph is acyclic, the arc between v_i and v_j must be from the one with the smaller number to the one with the larger. This fixes the indegree and outdegree also at the required values. But the indegree and outdegree are not dependent on the spanning path or the logical numbering. Hence both spanning path and logical numbering are unique. ■

This relationship between the outdegrees, and between the indegrees (either directly implies the other) can be used to prove a tournament to be acyclic.

Theorem 7.14

Let T be a tournament with p vertices, and suppose that the outdegrees of the vertices are

$$p-1, p-2, \ldots 2, 1, 0.$$

Then T is acyclic.

Proof

Let the vertices be numbered in such a way that v_i has outdegree $p - i$, and therefore indegree $i - 1$. Consider v_1. It has indegree 0, so for all k, the arcs incident with v_1 are $v_1 v_k$. In particular, $v_1 v_2$ is an arc. As v_2 has indegree 1, this is the only arc into v_2. Hence the arc between v_2 and v_3 is $v_2 v_3$. The two arcs into v_3 are $v_1 v_3$ and $v_2 v_3$ so the arc between v_3 and v_4 is $v_3 v_4$. And so on. This may be converted into a formal inductive proof using the subtournament defined by $V \setminus \{v_1, v_2, \ldots, v_m\}, m = 1, 2, \ldots, p$.

7.6.3. Outdegree Analysis

The previous section has shown that an acyclic tournament can be recognised from the list of outdegrees (or equivalently from the indegrees). In this section we show that the whole component structure of a general tournament can be obtained from the list of outdegrees. We shall reduce the process to a mechanical calculation. Suppose that the components are $C_1, C_2, \ldots C_k$ in non-increasing order of cardinality and that the cardinalities of these components are $c_1, c_2, \ldots, c_k$, respectively.

Theorem 7.15

If T is a tournament, and u is a vertex in component C_i, of T, then

$$c_{i+1} + c_{i+2} + \ldots + c_k \leq od(u) \leq c_i + c_{i+1} + \ldots + c_k - 1, \qquad \text{and}$$
$$c_1 + c_2 + \ldots + c_{i-1} \leq id(u) \leq c_1 + c_2 + \ldots + c_i - 1.$$

Proof

Every member of $C_i \cup C_{i+1} \cup \ldots \cup C_k$ is a successor of u. On the other hand, at least one member of C_i, namely u itself, is not a successor of u. (And if $C_i \neq \{u\}$ there will be another also.) The proof for indegrees is similar. ∎

Theorem 7.16

If u and v are vertices in a tournament and $u \in C_i$ but $v \in C_j$ with $i < j$ then

$$od(u) > od(v), \text{ and}$$
$$id(u) < id(v).$$

Proof

Omitted.

Theorem 7.17

If u and v are vertices in a tournament, and $od(u) = od(v)$, then u and v belong to the same component.

Proof

Omitted.

Suppose now we partition the vertices of a tournament into two nonempty disjoint sets M and N. That is:

$$M \cup N = V \text{ and } M \cap N = \emptyset.$$

Let M and N have m and n vertices respectively. Consider the sum of the outdegrees of the vertices in M. This is the number of arcs which begin and end in M. There are $\frac{1}{2}m(m-1)$ such arcs. Then there are the arcs which begin in M and end in N. There may be no such arcs or as many as mn, the latter occurring only if every member of M is a precursor of every member of N. Thus we have

$$\tfrac{1}{2}m(m-1) \leq \sum_{u \in M} od(u) \leq \tfrac{1}{2}m(m+2n-1).$$

Theorem 7.18

Let T be a tournament on p vertices and let V be its vertex set. If $V = M \cup N$, where $M \cap N = \emptyset$, then $M = C_1 \cup C_2 \cup \ldots \cup C_i$ and $N = C_{i+1} \cup C_{i+2} \cup \ldots \cup C_k$, for some i, if and only if the sum of the outdegrees of vertices in M is $\frac{1}{2}m(2p-m-1)$, where $C_1, C_2, \ldots, C_k$ are the components in order.

Proof

We saw informally before that the sum of the outdegrees is at most $\frac{1}{2}m(m+ 2n-1)$. But $m+n=p$, so this expression equals $\frac{1}{2}m(2p-m-1)$. This bound will be attained precisely when each arc between M and N has its beginning in M and its end in N. This will happen when

$$M = C_1 \cup C_2 \cup \ldots C_i$$

and, for some i,

$$N = C_{i+1} \cup C_{i+2} \cup \ldots \cup C_k.$$

For if M and N are as above, and every arc between them begins in M and ends in N, and if every arc between them is from M to N, then no component has members in both M and N and further, every component contained in M precedes every component in N. ■

This result shows us that if we collect together sets of vertices such that their outdegree sums have the right value, this will enable us to break down the tournament into components. Fortunately we do not have to proceed by hit-and-miss methods.

Theorem 7.16 implies that if we list the vertices in order of decreasing outdegree, we group the vertices in the various components together. It is then sufficient to test the first m members of the list for each m, add their outdegrees and compare the sum with $\frac{1}{2}m(2p-m-1)$, which can be computed in advance. In fact we go a little further in order to simplify the computations.

Suppose the vertices in non-increasing order of outdegree are $u_1, u_2, \ldots, u_p$. Then we have to compare:

$$od(u_1) + od(u_2) + \ldots + od(u_m)$$

with

$$\tfrac{1}{2}m(2p-m-1).$$

Now

$$\tfrac{1}{2}m(2p-m-1) = (p-1) + (p-2) + \ldots + (p-m)$$

So u_m is the last vertex in the list in its component if and only if

$$od(u_1) + od(u_2) + \ldots + od(u_m)$$
$$= (p-1) + (p-2) + \ldots + (p-m).$$

That is

$$\{(p-1) - od(u_1)\} + \{(p-2) - od(u_2)\} + \ldots + \\ \{(p-m) - od(u_m)\} = 0.$$

We call the term

$$(p - i) - od(u_i),$$

the *deficit* of u_i and write it as

$$def(u_i).$$

Hence the condition for u to be the last vertex in its component is that

$$cum(u_m) = def(u_1) + def(u_2) + \ldots + def(u_m) = 0, \qquad (7.5)$$

where $cum(u_m)$ is called the *cumulative deficit* at u_m. We observe from (7.5) that

$$cum(u_m) = cum(u_{m-1}) + def(u_m).$$

We construct a table to facilitate the calculation.

7.6.4 Tournament Wins Analysis

1. Construct a table with five columns, headed: *vertex, od, p − m, def,* and *cum*.

2. In column *vertex* list the vertices in non-increasing order of outdegree.

3. In column *od* list the corresponding outdegrees.

4. Let p be the number of vertices in the tournament. In column $p - m$ write $p - 1$ in the first row, decreasing by 1 at each row to 0 in the bottom row.

5. In each row subtract the entry under *od* from the entry under $p - m$ and enter the result in the column *def*.

6. Copy the first entry in column *def* into column *cum*. To obtain the remaining entries in the *cum* column add the entry to the left to the entry above.

7. Rule across the table under every 0 in the *cum* column. As a check, the final entry must be 0, and none can be negative.

8. The lines partition the vertices into the components.

As an example, we analyse the tournament of Figure 7.4 in Table 7.1.

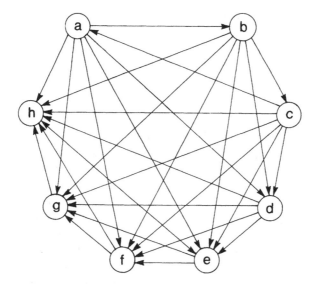

Figure 7.4 Tournament wins analysis

Vertex	od	p − m	def	cum
a	6	7	1	1
b	6	6	0	1
c	6	5	−1	0
d	4	4	0	0
e	2	3	1	1
f	2	2	0	1
g	1	1	0	1
h	1	0	−1	0

Table 7.1 A tournament wins analysis.

The components are therefore

$$\{a, b, c\}, \{d\}, \{e, f, g, h\}.$$

From the list of components we can reconstruct most of the individual arcs. To be precise, we know the direction of all arcs whose beginning and end are in different components. If we are given a little more information we can reconstruct the remaining arcs by a similar chain of reasoning.

In the present case, for example, suppose that we are given the components, outdegrees and the facts that ab and he are arcs. Within the component $\{a, b, c\}$, each vertex is the beginning of one arc and the end of one arc. If ab is an arc, therefore, so are bc and ca. Within $\{e, f, g, h\}$ the argument is a little more complex. As he is an arc and h has outdegree 1, the other arcs at h are fh and gh. As e is the end of one arc within the subtournament defined by $\{e, f, g, h\}$, ef and eg are arcs. This leaves the arc between f and g, which must be fg to balance the outdegrees.

This kind of argument can be used generally. Each component must be treated separately, and, in general, the number of arcs which need to be given before a complete solution can be obtained rises with the number of vertices in the component.

7.7 Summary

In this chapter we have discussed some of the important properties of digraphs. There are many parallels with graphs. However, there are sufficient properties which are not possessed by graphs to make digraphs of theoretical interest in their own right.

To this end we have studied digraph connectivity, traversability, trees, matrices, and tournaments. As will be seen in Part II of this book, digraphs are of utility in the solution of many practical problems.

7.8 Exercises

7.1 Suppose that four judges: J_1, J_2, J_3, and J_4 each rank eight objects: $O_1, O_2, \ldots, O_8$ independently. Their rankings are

Judge	Rankings							
	1st	2nd	3rd	4th	5th	6th	7th	8th
$J1$	O_1	O_2	O_3	O_4	O_5	O_6	O_7	O_8
$J2$	O_2	O_4	O_6	O_8	O_1	O_3	O_5	O_7
$J3$	O_3	O_5	O_4	O_8	O_7	O_6	O_2	O_1
$J4$	O_6	O_7	O_1	O_2	O_3	O_4	O_5	O_8

Construct a digraph which reflects these rankings. Use component analysis to interpret these rankings.

7.2 Show that the converse of a strong digraph is strong.

7.3 Is every subdigraph of a strong digraph itself strong?

7.4 If D is a strong digraph with n vertices, what is the least number of arcs in D?

7.5 If D is a digraph with n vertices, that is not strong, what is the maximum number of arcs in D?

7.6 Suppose that a tournament has nine vertices, two components, and that exactly half of its arcs join vertices in the same component. How many vertices are in each component?

7.7 Show that D^*, the condensation of any digraph D, is acyclic.

7.8 Let u and v be arcs in a digraph D, with a walk from u to v. If C_u and C_v are the components containing u and v respectively, show that there is a walk from C_u to C_v in D^*.

7.9 Using the notation of Exercise 7.8, show that, if there is a walk from C_u to C_v in D^*, then there is a walk from u to v in D.

7.10 Show that if D^* is strong then D^* comprises solely an isolated vertex.

7.11 Show that D^* comprises a unique spanning walk, and is unilateral, if and only if D is unilateral.

7.12 Find examples of each of the following:
   (a) A unilateral digraph which has a subdigraph which is not unilateral.
   (b) A digraph which is not unilateral which has a non-trivial subdigraph which is unilateral.
   (c) A unilateral digraph which does not have a spanning path.

7.13 Prove that the converse of a unilateral digraph is unilateral.

7.14 Prove that every digraph with a unilateral partial digraph is unilateral.

7.15 Prove that the converse of a tournament is a tournament.

7.16 Prove that if a tournament is not acyclic then there is at least one pair of vertices with the same outdegree.

7.17 Prove all observations, and Theorems 7.16 and 7.17, of this chapter.

7.18 Count all vertex-labelled digraphs with no more than four vertices.

7.19 Identify all oriented graphs with no more than three vertices.

7.20 Prove that every arc in a digraph belongs to either a cycle or a directed cut-set.

7.21 Prove that every Eulerian digraph is strongly connected. Is the converse true?

7.22 Illustrate relationships (7.3), (7.4) for the digraph in Figure 6.2(b).

7.23 Prove that every strong tournament with n vertices has an m-cycle for $m = 3, 4, \ldots n.$

7.24 Prove that if a digraph with distinct vertices v_1 and v_2, contains a path from v_1 to v_2 and a directed path from v_2 to v_1 then it is cyclic.

8 Coverings and Colourings

The purest and most thoughtful minds are those which love colour the most.

The Stones of Venice II, v — John Ruskin

A graph theoretic concept of interest, both from the theoretical and practical points of view, is that of *covering*. A vertex (edge) of a graph is said to *cover* the edges (vertices) with which it is incident.

Thus it is natural to ask, for a given graph G, what is the minimum number of edges (vertices) needed to cover all the vertices (edges) of G? One can also ask a converse question. For a given graph G, what is the maximum number of edges (vertices) which are mutually nonadjacent? Such sets of edges (vertices) are termed *independent*.

As well as calculating the four numbers just mentioned, and their corresponding edge and vertex sets, one can also identify minimal (or maximal) sets of edges, or vertices which cover all the vertices or edges of a given graph.

This notion of independence can be extended to a more general one of dominance. A set of edges (vertices) in a graph G, is said to be *dominant* if every edge (vertex) of G either belongs to the set or is adjacent to a member of it.

In Section 1.1.3 we introduced the Four Colour Theorem, which establishes that it is possible to colour the regions of any map drawn in a plane so that every pair of adjacent regions have different colours. One can view the colouring of such maps in graph theoretic terms by creating a (planar) graph for any planar map. This graph will have a vertex for every region of the map, and a pair of its vertices are directly joined by an edge whenever the regions they represent in the map have a common boundary. The colouring of a region of the map is equivalent to associating a colour with the corresponding vertex of the graph. A graph constructed in this way will have the same colouring structure as the map it represents. Later in this chapter we shall consider the vertex colourings of graphs and chromatic numbers.

We now introduce a concept closely related to that of edge independence, namely *matching*. As with the other topics of this chapter, matching is of great utility. This is because in many practical problems, one is often endeavouring to assign one set of objects to another set, in a one-to-one fashion, such as workers to machines or children to schools. Such an assignment is often called a *matching* and can be modelled in graph theoretic terms, as shown later in this chapter.

The topics of this chapter: covering, colouring, and matching are of practical importance in operations research, as will be seen in Chapter 12. We begin with an introduction to covering.

8.1 Covering, Independence, and Domination

Definitions

An edge $\{u, v\}$ in a graph is said to *cover* its incident vertices u and v.

A vertex in a graph is said to *cover* the edges with which it is incident.

If $G = (V, E)$ is a graph and $E' \subseteq E$, then E' is said to be an *edge cover* of G and to *cover* G if for each vertex $v \in V$ there is at least one edge in E' which covers v.

If $G = (V, E)$ is a graph and $U \subseteq V$, then U is said to be a *vertex cover* of G and to *cover* G if for each edge $e \in E$, there is at least one vertex in U which covers e.

For a given graph G the cardinality of the edge cover with the least number of elements is called the *edge covering number* of G and is denoted by $\alpha_1(G)$ or α_1.

For a given graph G the cardinality of the vertex cover with the least number of elements is called the *vertex covering number* of G and is denoted by $\alpha_0(G)$ or α_0.

A covering is said to be *minimal* if none of its proper subsets is a covering.

A covering in a graph G is said to be *minimum* if there is no covering of G with a smaller number of elements.

These ideas can be illustrated via the graph in Figure 8.1(a). The set $U = \{v_1, v_2, v_3, v_4\}$ is a minimal vertex cover in the sense that it covers all the edges of the graph but no proper subset of it does so. Yet U does not correspond to the vertex covering number because $U_1 = \{v_5\}$ is the unique minimum vertex cover and thus $\alpha_0 = 1$. Also $\alpha_1 = 4$ as the set of edges $\{e_1, e_2, e_3, e_4\}$ is the minimum edge cover.

More generally, the edges (vertices) of any spanning tree, Hamiltonian path or unicursal path of any connected graph G, constitute an edge (vertex) cover of G.

Let G be a graph with vertex set V. We can make some observations about edge coverings in G.

C(i)   An edge covering of G can always be found so long as G does not contain an isolated vertex.

C(ii)   If $|V| = n, where(n > 1)$, then any edge covering of G will contain at least $n/2$ edges. If $G = K_n$, then $\alpha_1 = [(n+1)/2]$, the integer part of $(n+1)/2$.

C(iii)   Every edge covering includes every pendant edge.

C(iv)   It is possible to remove a subset of edges (possibly empty) from any edge covering of G in order to create a minimal (but not necessarily minimum) edge covering of G.

C(v)   Minimal edge coverings are acyclic.
         We can make similar observations about vertex coverings:

C(vi)   A vertex covering exists for any graph G.

C(vii)   If $G = K_n$, then $\alpha_0 = n - 1$.
          A vertex covering may have only a single element. [This is true for any *star*, a graph which possesses a unique vertex (called the *centre*) with which every edge is incident. Figure 8.1(a) depicts a star with centre v_5.]

C(viii)   It is possible to remove a subset of vertices (possibly empty) from any vertex covering in order to create a minimal (but not necessarily minimum) vertex covering.

We now prove a simple theorem concerning edge coverings.

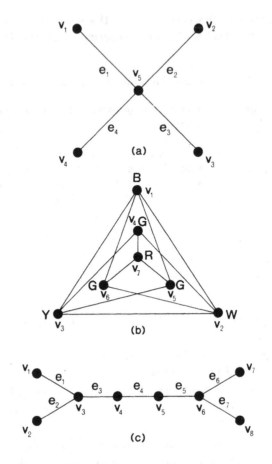

Figure 8.1 (a) Cover, independence and dominance
(b) colouring (c) matching

Theorem 8.1

An edge covering in a graph does not contain a path of at least three edges
if and only if it is minimal.

Proof

Consider an edge covering containing a path of at least three edges. The
second edge of the path can be removed, leaving an edge covering. Thus
the original covering is not minimal.

Suppose that there exists an edge covering which does not contain a path of at least three edges. In this case, each component of the covering is a star. Because it is impossible to remove an edge from an edge covering of a star (see observation C(vii)), the edge covering is minimal. ∎

We turn now to the second concept introduced at the beginning of this chapter, namely *independence*.

Definitions

If $G = (V, E)$ is a graph and $E' \subseteq E$, then E' is said to be *edge independent* if no two edges of E' are adjacent.

If $G = (V, E)$ is a graph and $U \subseteq V$, then U is said to be *vertex independent* if no two vertices of U are adjacent.

For a given graph G, the cardinality of the set of edges of G which is the largest independent set of G is called the *edge independence number* of G and is denoted by $\beta_1(G)$ or β_1.

For a given graph G, the cardinality of the set of vertices of G which is the largest vertex - independent set of G, is called the *vertex independence number* of G, and is denoted by $\beta_0(G)$ or β_0.

An independent set is said to be *maximal* if none of its proper supersets is independent.

An independent set in a graph G is said to be *maximum* if there is no independent set in G with a greater number of elements.

We illustrate these ideas with the graph in Figure 8.1(a). Any one of the edges of the graph constitutes an edge - independent set. Such a singleton set is at once maximal and maximum.

The set $\{v_5\}$ is a maximal vertex - independent set. However it is not maximum because of the existence of the $\{v_1, v_2, v_3, v_4\}$. Thus in this graph, $\beta_0 = 4$ and $\beta_1 = 1$.

We can make some observations about edge - independent sets in any graph G.

I(i)   An edge - independent set can always be found if G contains at least one edge. Any single edge of G constitutes such a set.

I(ii)   If $G = K_n$, the complete graph on n vertices, then $\beta_1 = [n/2]$, the integer part of $n/2$.

I(iii)   Every edge in G belongs to at least one maximal edge - independent set.

I(iv)   It is possible to add a subset of edges (possibly empty) to any edge - independent set in G in order to create a maximal (but not necessarily maximum) edge - independent set in G.

We can make similar observations about vertex - independent sets in G.

I(v)   A vertex - independent set exists for G. (Any single vertex of G constitutes such a set).

I(vi)   If $G = K_n$, then $\beta_0 = 1$.

I(vii)   Every vertex in G belongs to at least one maximal vertex - independent set.

I(viii)   It is possible to add a subset of vertices (possibly empty) to any vertex - independent set in G in order to create a maximal (but not necessarily maximum) vertex - independent set in G.

We now state theorems linking the notions of covering and independence.

Theorem 8.2

For any connected graph G, with n vertices, and e edges in which $n+e > 1$,

$$\alpha_0(G) + \beta_0(G) = \alpha_1(G) + \beta_1(G) = n.$$

Proof (Gallai (1959).)

Let G be a connected graph with at least one edge.

(i) To show that

$$\alpha_1(G) + \beta_1(G) = n.$$

Let E_1 be an edge - independent set of β_1 edges in G. An edge cover of G, E' say, can be constructed as E_1 together with an additional edge for each vertex of G not covered by any edge in E_1. Because

$$|E_1| + |E'| \leq n,$$

and

$$|E'| \geq \alpha_1,$$

we have

$$\alpha_1 + \beta_1 \leq n.$$

To establish the reverse inequality, let $\overline{E}$ be a minimum edge cover of G. By definition, $\overline{E}$ does not possess an edge whose two incident vertices are incident with edges in $\overline{E}$. Thus $\overline{E}$ comprises a set of subsets of edges of G where each subset constitutes a star. A set of edges, say $\overline{E}$ comprising exactly one edge from each star, is edge - independent.

But

$$|E'| + |\overline{E}| = n,$$

and

$$|\overline{E}| \leq \beta_1.$$

Thus

$$\alpha_1 + \beta_1 \geq n.$$

(ii) To show that

$$\alpha_0 + \beta_0 = n.$$

Let V_1 be any maximum independent set of β_0 vertices of G. Because no edge in G joins two vertices of V_1, the $n - \beta_0$ remaining vertices of G constitute a vertex cover of G. Hence

$$\alpha_0 \leq n - \beta_0.$$

However if V' is a minimum vertex cover of G, then no edge can join any two of the $n - \alpha_0$ vertices of G. Therefore $V \backslash V'$ is independent. Thus

$$\beta_0 \geq n - \alpha_0.$$

■

Theorem 8.3

If G is a bipartite graph then

$$\alpha_0(G) = \beta_1(G).$$

Proof

The reader is referred to the proof given by König (1931).

A further notion introduced earlier was that of dominance.

Definitions

An edge (vertex) in a graph G is said to *dominate* those other edges (vertices) in G with which it is adjacent.

If $G = (V, E)$ is a graph and $E_1 \subseteq E$, $(U \subset V)$ then E_1 (U) is said to be an *edge (vertex) dominating set* for G if every edge of E (vertex of V) either belongs to E_1 (U) or is dominated by an edge of E_1 (vertex of U).

For a given graph $G = (V, E)$, the cardinality of the edge (vertex) - dominating set with the least number of elements is called the *edge (vertex) dominating number* of G and is denoted by $\sigma_1(G)$ or σ_1 (by $\sigma_0(G)$ or σ_0).

A dominating set is said to be *minimal* if none of its proper subsets are dominating.

A dominating set of a graph G is said to be *minimum* if there is no dominating set of G with a smaller number of elements.

Let us study the graph in Figure 8.1 (a) in order to illustrate the notion of dominance. Any one of the edges of the graph constitutes an edge - dominating set, which is at once minimal and minimum. The set $\{v_1, v_2, v_3, v_4\}$ is a minimal vertex - dominating set. However it is not minimum as $\{v_5\}$ is dominating. Thus $\sigma_1 = \sigma_0 = 1$.

More generally, the edges (vertices) of any spanning tree, Hamiltonian path, or unicursal path of any connected graph G, constitute an edge (vertex) - dominating set of G.

Let $G = (V, E)$ be a graph. We can make some observations about edge - dominating sets for G:

D(i)   An edge - dominating set can always be found if G contains at least one edge.

D(ii)   If $G = K_n$, then $\sigma_1 = \lceil n/2 \rceil$.

D(iii)   A vertex - independent set of G is a vertex - dominant set for G only if it is maximal.

D(iv)   It is possible to remove a subset of edges (possibly empty) from any edge - dominating set for G in order to create a minimal (but not necessarily minimum) edge - dominating set of G.

D(v)   A minimal edge - dominating set for G is not necessarily an edge - independent set of G.

D(vi)   Every maximal edge - independent set of G is an edge - dominating set for G. (See the proof of Theorem 8.4).

D(vii)   An edge - independent set of G is an edge - dominant set for G only if it is maximal.

We can make similar observations about vertex - dominating sets for G.

D(viii)   A vertex - dominating set for G exists.

D(ix)   If G is complete, then $\sigma_0 = 1$.

D(x)   If G is connected with at least one internal vertex, then every minimum vertex - dominating set for G does not contain any pendant vertices.

D(xi)   It is possible to remove a subset of vertices (possibly empty) from any vertex - dominating set for G in order to create a minimal (but not necessarily minimum) vertex - dominating set for G.

D(xii)   A minimal vertex - dominating set for G is not necessarily a vertex - independent set of G.

D(xiii)   Every maximal vertex - independent set of G is a vertex - dominating set for G. (This is established in most proofs of Theorem 8.5.)

We now state two theorems linking the notions of independence and dominance.

Theorem 8.4

For every graph $G = (V, E)$ with at least one edge,

$$\sigma_1(G) \le \beta_1(G).$$

Proof

Let E_1 be a maximum edge - independent set of G. By assumption and observation I(i), E_1 can always be found. By definition, $|E_1| = \beta_1(G)$.

Every edge $e \in E \backslash E_1$, is adjacent to some edge of E_1, otherwise $E_1 \cup \{e\}$ is an independent set of edges with cardinality $\beta_1 + 1$. This is impossible as E_1 is assumed maximum. Hence E_1 is an edge - dominating set for G. Hence the result. ■

Theorem 8.5

For every graph $G = (V, E)$,

$$\sigma_0(G) \le \beta_0(G).$$

Proof

Analogous to that of Theorem 8.4.   ■

We turn now to graph vertex colouring.

8.2 Colouring

Definition

A *colouring* of a graph G, is an assignment of colours to its vertices so that no two adjacent vertices in G have the same colour.

We can see from this definition that a (vertex) colouring of a graph is closely allied with concepts introduced in the previous section. Since a colouring requires a partitioning of the set of vertices of a graph into non-adjacent subsets, it has similarities with the identification of a vertex - independent set. Figure 8.1 (b) shows a colouring of a graph. Studying this colouring prompts the obvious question, can the vertices be re-assigned colours so that less than five colours can be used in a colouring of this graph? Because the graph is planar, by the Four Colour Theorem (Section 1.1.3), we know that it is possible to find a colouring with four colours. But is four the minimum number of colours that can be used? Indeed what is the minimum number of colours that must be used in any colouring of this graph? It turns out that four is indeed the minimum number of colours required, and is termed the *chromatic number* of the graph. This leads us to some more definitions:

Definitions

The set of vertices assigned any one colour in a colouring of a graph is termed a *colour class*.

(By definition, a colour class in a vertex-coloured graph is a vertex - independent set.)

A colouring of a graph which assigns c colours to the vertices of a graph is termed a *c-colouring*.

(Thus a c-colouring partitions the vertices of a graph into c colour classes.)

The *chromatic number* of a graph G is the minimum number χ, for which G has a χ - colouring, and is denoted by $\chi(G)$ or χ.

A graph G is *c-colourable* if $\chi(G) \leq c$ and *c-chromatic* if $\chi(G) = c$.

Thus the colouring of the graph in Figure 8.1(b) demonstrates that it is 5-colourable. Actually it is 4-chromatic.

For obvious reasons we consider only connected graphs in studying colouring. Let $G = (V, E)$ be a connected graph. We can make some observations about any colouring of G.

VC(i) G is 1-chromatic if and only if it has no edges.

VC(ii)   G is c-chromatic, where $c > 1$, if and only if G has at least one edge.

VC(iii) If $G = K_n$, then G is n-chromatic.

VC(iv)   If G contains K_n as a subgraph then G is c-chromatic, where $c \geq n$.

VC(v)   If G is an n-cycle then G is 3-chromatic if n is odd and 2-chromatic if n is even.

VC(vi) G is bipartite if and only if G is 2-colourable.

VC(vii) If G is a tree with at least one edge then G is 2-chromatic.

VC(viii) G is 2-colourable if and only if G has no cycles with an odd number of edges.

VC(ix)   The chromatic number of G is at most one greater than the degree, say d_{max}, of its vertex with maximum degree.

VC(x)   If G does not have a complete subgraph with $(d_{max} + 1)$ vertices then the chromatic number of G is at most d_{max}.

VC(xi)   A colouring of G partitions V into vertex - independent sets.

Following on from the last observation we have:

Theorem 8.6

If $G = (V, E)$ is a graph with n vertices and chromatic number χ then $\chi \beta_0(G) \geq n$.

Proof

This follows easily from the fact that in any χ-colouring of G the largest number of vertices in G with the same colour cannot exceed the vertex independence number $\beta_0(G)$, of G. ∎

We now raise the more practical matter of actually finding the chromatic number of a given graph $G = (V, E)$. We could do this by finding the minimum number of maximal vertex - independent sets which together form a partition of V. For the graph in Figure 8.1(a) this partition is: $\{v_1, v_2, v_3, v_4\}$ and $\{v_5\}$. The number of sets equals the chromatic number of the graph and thus $\chi = 2$. For the graph in Figure 8.1 (b) such a partition is $\{v_1, v_4\}$, $\{v_2, v_5\}$, $\{v_3, v_6\}$ and $\{v_7\}$. Thus, in this case, $\chi = 4$. This analysis leads us to what is known as the *chromatic partitioning problem*:

> **Given a connected graph $G = (V,E)$, partition V into the minimum possible number of colour classes.**

By definition, each colour class corresponds to a maximal vertex - independent set. Thus the vertex sets we have just listed for the graphs in

Figures 8.1 (a) and (b) constitute chromatic partitionings for those graphs. The chromatic partition given for the graph in Figure 8.1 (a) is unique. The complete list of chromatic partitionings for the graph in Figure 8.1 (b) is:

$$\{v_1, v_4\} \quad \{v_2, v_5\} \quad \{v_3, v_6\} \quad \{v_7\}$$
$$\{v_1, v_7\} \quad \{v_2, v_5\} \quad \{v_3, v_6\} \quad \{v_4\}$$
$$\{v_1, v_4\} \quad \{v_2, v_7\} \quad \{v_3, v_6\} \quad \{v_5\}$$
$$\{v_1, v_4\} \quad \{v_2, v_5\} \quad \{v_3, v_7\} \quad \{v_6\}$$
$$\{v_4, v_5\} \quad \{v_2, v_7\} \quad \{v_3, v_6\} \quad \{v_1\}$$
$$\{v_4, v_6\} \quad \{v_2, v_5\} \quad \{v_3, v_7\} \quad \{v_1\}$$
$$\{v_1, v_7\} \quad \{v_4, v_5\} \quad \{v_3, v_6\} \quad \{v_2\}$$
$$\{v_1, v_4\} \quad \{v_5, v_6\} \quad \{v_3, v_7\} \quad \{v_2\}$$
$$\{v_1, v_7\} \quad \{v_2, v_5\} \quad \{v_4, v_6\} \quad \{v_3\}$$
$$\{v_1, v_4\} \quad \{v_2, v_7\} \quad \{v_5, v_6\} \quad \{v_3\}$$

Thus there are many 4-colourings of this graph. A graph for which there is only a single χ-colouring is termed *uniquely colourable*. An example of a uniquely colourable graph is given in Figure 8.1 (a). We shall discuss graph colouring from an algorithmic point of view in Section 9.6.2.

8.3 Matching

Definition

If $G = (V, E)$ is a graph and $E_1 \subseteq E$, then E_1 is called a *matching* if E_1 is edge-independent.

The reason for the term *matching* in the above definition is that the two vertices incident with an edge e, in E_1 are associated, paired, or *matched* by e.

Definitions

A matching in a graph is said to be *maximal* if it is maximal edge- independent.

A matching in a graph is said to be *maximum* if it is maximum edge-independent.

As an example, consider the graph in Figure 8.1 (c). $M = \{e_3, e_5\}$ is edge-independent. It matches the vertices: v_3 and v_4, and also v_5 and v_6. However M is also maximal, as no further edge can be added to it without destroying its edge-independence. However M is not maximum, a property possessed by the set $\{e_1, e_4, e_6\}$.

The reader may have noticed a number of facts about the graph in Figure 8.1 (c). Firstly it is bipartite, with vertex partition: $\{v_1, v_2, v_4, v_6\}$ and $\{v_3, v_5, v_7, v_8\}$. Secondly, the number of edges in any maximum matching is 3, which equals the vertex covering number, i.e. $\alpha_0 = \beta_1$. This is true in general for any bipartite graph, as stated in Theorem 8.3.

In order to characterize maximum matchings, we introduce some more definitions:

Definitions

Let G be a graph with a matching $M \subseteq E$, and a path P. Then P is termed *alternating* (with respect to M) if its edges are alternately in M and not in M.

An alternating path P, with respect to a matching M is termed *augmenting* if its end vertices are not incident with any edge of M.

A matching M, in a graph G is said to be *unaugmentable* if G does not have an augmenting path with respect to M.

This leads us to the following theorem:

Theorem 8.7

A matching in a graph G is unaugmentable if and only if it is maximum.

Proof (Berge (1957).)

(Sufficiency)
Obvious.

(Necessity)
Let M be an unaugmentable matching of G. Let M' be a maximum matching with the property that, among all maximum matchings $\overline{M}$, of G, the number of edges, say p, in $M \backslash \overline{M}$ is a minimum. If $p = 0$ then $M = M'$. If $p > 0$, we identify a trail, say T, with the maximum possible number of edges, whose edges are alternatively in $M \backslash M'$ and in M'. Because M' is unaugmentable, T cannot have its beginning and ending edges in $M \backslash M'$.

Also T has the same number of edges in M as in $M\backslash M'$. Construct a maximum matching from M by substituting those edges of T which are in M' by the edges of T which are in $M\backslash M'$. This implies that $M\backslash\overline{M}$ has fewer edges than $M\backslash M'$. However this is impossible, given the definition of M'. ∎

On looking at the maximum matching $\{e_2, e_4, e_7\}$, in the graph in Figure 8.1 (c), it is clear that it does not match vertex v_1, nor vertex v_7. Of course while it is possible to find a matching which matches any vertex in this graph, a matching which matches all of its vertices does not exist. This prompts the question, when does a graph possess a matching which matches all of its vertices?

We start to examine this question with the following definition.

Definition

A matching M, in a graph G is said to be *perfect*, or a *1-factor*, or a *dimer covering*, if some edge of M is incident with every vertex of G.

A problem involving dimer coverings will be discussed in Section 15.1.1. A necessary and sufficient condition for a graph to have a perfect matching was first reported by W. Tutte (1947). Before stating the theorem, we need some more notation.

A subgraph of a graph $G = (V, E)$ obtained from G by deleting the vertices in U together with their incident edges, is denoted by $G - U$.

Also the set of components of a graph G, having an odd number of vertices is denoted by $o(G)$.

Theorem 8.8

A graph $G = (V, E)$ has a perfect matching if and only if $o(G - U) \leq |U|, \forall U \subseteq V$.

Proof

The reader is referred to the proof given by Lovasz (1975).

Despite the fact that it has an even number of vertices (an obvious necessary condition for the existence of a perfect matching) the graph in Figure 8.2 (a) does not have a perfect matching. However the graph in Figure 8.2 (b) does have a perfect matching, as indicated by the solid lines.

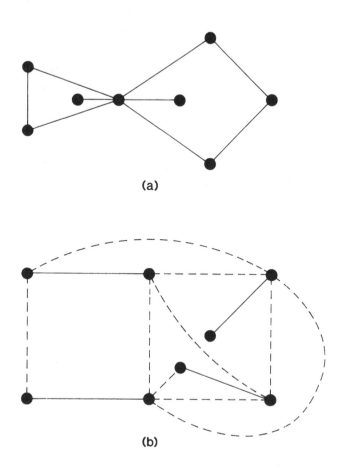

(a)

(b)

Figure 8.2 (a) A graph which does not contain a perfect matching
(b) a perfect matching

Because many of the matching problems in operations research can be
modelled by bipartite graphs, we state results for that special case. We
need a definition first.

Definition

A matching M, in a bipartite graph $G = (V, E)$, with bipartite vertex
partition V_1 and V_2, is said to be *complete in V_1* if there is one edge of M
incident with every vertex of V_1.

This means that a matching M, in a *bipartite* graph $G = (V, E)$, with bipartite vertex partition V_1 and V_2, is complete in V_1 ($\subset V$) if every vertex in V_1 is matched by M (to a vertex in V_2).

A complete matching in $V_1 = \{v_1, v_2, v_3\}$ is shown as the solid lines in Figure 8.3 (a). Of course this matching is a *maximum* matching, as all complete matchings are. However there are maximal matchings which are not complete. One is shown by the solid lines in Figure 8.3 (b), where $V_1 = \{v_1, v_2, v_3\}$.

When does a bipartite graph possess a complete matching in V_1? Obviously a necessary (but not sufficient) condition is that $|V_1| \leq |V_2|$.

We now state a necessary and sufficient condition for a complete matching to be present.

Theorem 8.9

If G is a bipartite graph with bipartite vertex partition: V_1 and V_2, G has a matching that is complete in V_1 if and only if every subset of p vertices of V_1 is collectively adjacent to at least p vertices in V_2 for all $p = 1, 2, \ldots, |V_1|$.

Proof

(Sufficiency)
Obvious.

(Necessity)
By induction on p. ■

Let $V_1 = \{v_1, v_2, v_3\}$. It is instructive to apply Theorem 8.9 to establish that the graph in Figure 8.3(b) has a matching that is complete in V_1, but that the graph in Figure 8.3(c) does not. The former fact is established because each vertex of V_1 in Figure 8.3(b) is adjacent to at least two vertices, and all three vertices in V_1 are collectively adjacent to four vertices in V_2. The latter fact is established because, in Figure 8.3(c), vertices v_1 and v_3 are collectively adjacent to only a single vertex of V_2.

We can use such checking of collective adjacency to establish whether a given bipartite graph has a complete matching or not. Unfortunately, such checking is very laborious for large, general bipartite graphs.

The following corollary establishes whether a complete matching exists.

Corollary 8.1

If G is a bipartite graph, with bipartite vertex partition V_1 and V_2, and

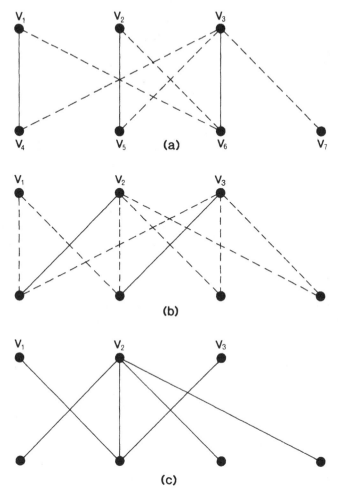

Figure 8.3   (a) A complete matching   (b) a maximal but not
complete matching
(c) a graph without a complete matching

there exists an integer p such that

$$\min_{v \in V_1}\{d(v)\} \ge p \ge \max_{v \in V_2}\{d(v)\},$$

then G has a complete matching in V_1.

Proof

Let $U \subseteq V_1$ where $|U| = q$. By assumption these q vertices are incident with at least pq edges. Because G is bipartite, each of the pq edges is incident with a vertex in V_2. However by assumption, the degree of each vertex in V_2 is at most p. Thus these pq edges are incident with at least q vertices in V_2. Thus the q vertices in U are collectively adjacent to at least q vertices in V_2. The result follows from Theorem 8.9. ■

Although Corollary 8.1 provides a sufficient condition for a bipartite graph to have a complete matching, it is not necessary. The graph in Figure 8.4(a) does not obey the condition of Corollary 8.1 where $V_1 = \{v_1, v_2, v_3\}$. Yet it has a complete matching in V_1, shown by the solid lines.

If a bipartite graph does not have a complete matching in certain applications (such as the assignment of jobs to workers) it is sometimes necessary to find a maximal matching. To this end we introduce a new definition.

Definition

If G is a bipartite graph with bipartite vertex partition: V_1 and V_2, then the *deficiency* of G, written $\delta(G)$ or δ, is defined as

$$\delta(G) = \max\{p - q\},$$

where the maximum is taken over all subsets of p vertices of V_1 which are collectively incident on a subset of q vertices of V_2.

We can restate Theorem 8.9 as

Theorem 8.10

If G is a bipartite graph with bipartite vertex partition V_1 and V_2, G has a matching that is complete in V_1 if and only if $\delta(G) \leq 0$.

Proof

As for Theorem 8.9. ■

This last theorem can be used to check to see whether or not a bipartite graph has a complete matching. The reader should establish that the graph in Figure 8.3 (c) has $\delta = 1$ (>0) and thus do not have a complete matching. The reason for choosing the term *deficiency* for δ is apparent from the next theorem.

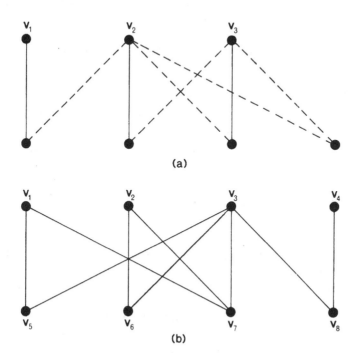

Figure 8.4 (a) A counter example to the necessity condition
            in Theorem 8.9

            (b) A graph used to illustrate the relationship
                between a matching and the adjacency matrix
                of the graph

Theorem 8.11

If G is a bipartite graph with bipartite vertex partition: V_1 and V_2, such
that $\delta(G) \geq 0$, then the maximum number of vertices in V_1 that can be
matched by any matching of G is $|V_1| - \delta(G)$.

Proof

The reader is referred to the proof given by Liu (1968).

Application of Theorem 8.11 to the graph in Figure 8.3 (c) implies that any
maximum matching contains exactly two edges. Recall that the adjacency

matrix of a graph was introduced in Section 6.1. We now explore the relationship between a matching and the adjacency matrix of a bipartite graph. Let G be a bipartite graph with bipartite vertex partition: V_1 and V_2. Suppose that $|V_1| = n_1$ and $|V_2| = n_2$. Then G is completely specified by the $n_1 \times n_2$ submatrix A_1, of the adjacency matrix of G, comprising the rows of A corresponding to the vertices of V_1 and the columns of A corresponding to the vertices of V_2. We can make some observations about A_1 with regard to any matching M_1 of G:

M(i)   There is a one-to-one correspondence between the edges of M and the unit entries of A_1.

M(ii)   No row or column of A_1 has more than one unit entry corresponding to an edge of M.

M(iii) M is complete in V_1 if there is exactly one unit entry in each row of A_1 corresponding to the edge of M.

M(iv) M is perfect if the unit entries of A_1 comprise an identity matrix within a permutation of rows.

M(v) M is maximal if no further unit entries of A_1 can be added to those unit entries of A_1 associated with the edges of M without creating a row or column of A_1 with at least two such entries.

We now illustrate these relationships via the graph in Figure 8.4 (b).

Let $V_1 = \{v_1, v_2, v_3, v_4\}$, and $V_2 = \{v_5, v_6, v_7, v_8\}$. Then

$$A_1 = \begin{array}{c} \\ v_1 \\ v_2 \\ v_3 \\ v_4 \end{array} \begin{array}{cccc} v_5 & v_6 & v_7 & v_8 \\ \left[\begin{array}{cccc} 1 & 0 & 1 & 0 \\ 0 & 1 & 1 & 0 \\ 1 & 1 & 1 & 1 \\ 0 & 0 & 0 & 1 \end{array}\right] \end{array}$$

If $M = \{v_1v_5, v_2v_6, v_3v_7, v_4v_8\}$, then M is complete, as M corresponds to the leading diagonal of A_1. Thus there is exactly one corresponding entry in each row of A_1. Because there is also exactly one corresponding entry in each column of A_1, M is perfect.

If $M = \{v_1v_7, v_3v_6, v_2v_8\}$, then M is maximal as no further edge can be added to M without one of the first three rows of A_1 or one of the last three columns of A_1 possessing two unit entries corresponding to the new M. We discuss matching from an algorithmic point of view in Section 9.6.3.

8.4 Summary

In this chapter we have studied the possibility of "covering" a graph in a number of ways. These include finding a subset of:

(i)   edges which cover the vertices of G,   (edge cover),
(ii) vertices which cover the edges of G,   (vertex cover),
(iii) edges which are independent,   (edge independence),
(iv) vertices which are independent,   (vertex independence),
(v)   edges which dominate the edges of G,   (edge dominance),
(vi) vertices which dominate the vertices of G, (vertex dominance),
(vii) vertices which colour the vertices of G, (vertex colouring), and
(viii) edges which match the vertices of G,   (vertex matching).

The developments of these concepts have much in common with each other. Many links between them have been stated by example, observation, and theorem. The concept of a perfect matching is of importance in crystal physics and will be discussed as the *dimer problem* in Section 15.1.1. As was mentioned earlier, some of the concepts of the present chapter are of interest in operations research and will be utilized in Chapter 12.

8.5 Exercises

8.1 Find minimum edge and vertex coverings of the graphs in Figures: 8.1 (b), (c), 8.2, 8.3, and 8.4.

8.2 Find minimum edge and vertex dominating sets for the graphs in Figures: 8.1 (b), (c), 8.2, 8.3, and 8.4.

8.3 Find a chromatic partitioning for the graphs of Figures 8.1 (a), (c), 8.2, 8.3, and 8.4.

8.4 Find vertex and edge - dominating sets, and a maximum matching in the graphs of Figures 8.1 (a), (b), (c), 8.3, and 8.4.

8.5 Prove the observations and theorems stated but not proved in this chapter.

8.6 Prove that the chromatic number of any graph G, cannot exceed the number of edges in the path with the most edges in G.

8.7 (The Marriage Problem) Suppose there is a finite set M, of single men and finite set W, of single women, and that each of the men knows a positive number of the women in W. Under what conditions can each man marry exactly one of the women that he knows?

8.8 Prove that a planar graph is a 2-colourable map if and only if each of its vertices is of even degree.

9 Algorithms

Though this be madness, yet there is method in it.

Hamlet II, ii — William Shakespeare

It is often of interest to analyse a graph in order to discover whether or not it possesses a certain property or to perform an optimization procedure upon the graph. Such an activity is often carried out by the use of an *algorithm* — a recipe for the solution of a given mathematical problem. We shall discuss some algorithms for identifying the properties of a given graph, which are useful in both combinatorics and in many application areas of graph theory. We shall also discuss optimization algorithms which are often used to find the subgraph of a given weighted graph which is optimal in some sense. Both of these types of algorithms are of importance in Part II of this book.

The first few sections of the present chapter establish a foundation upon which we can design efficient algorithms. A vital part of this foundation is the concept of computational complexity, which is a measure of the number of elementary computational steps required to implement a given algorithm. As computational complexity is a function of size of input, it is usually expressed in terms of the number of vertices and edges of an inputted graph or graphs. This raises the question of how to communicate the structure of a given graph to a computer. Also how should the computer manipulate the information with which it is provided? And how should it communicate the results of the implementation of the algorithm?

Questions must be asked, not only about the efficiency of a given algorithm, but also about its correctness, i.e. the guarantee that the algorithm will actually succeed in its stated aim. We begin with a discussion of algorithms.

9.1 Algorithms

An *algorithm* is a finite sequence of logical and mathematical instructions for the solution of a given, well-defined problem. The problem, the sequence of steps to be carried out, and the precise operations to be carried out at each step, must all be well-defined. For an algorithm to be of interest, it must possess, for every possible problem instance:

(i)   **Finiteness:**   Termination after a finite number of steps.

(ii) **Definiteness:**   Absence of ambiguity.

(iii) **Input:**   Specification of the information needed to begin.

(iv) **Output:**   Specification of the results of an implementation, and

(v)   **Effectiveness:** Guarantee of the resolution of any problem instance.

There are many ways in which an algorithm can be specified. These include:

(i) A carefully worded statement written in a normal language such as English,

(ii) A list of steps defined in mathematics and a normal language,

(iii) A flow chart,

(iv) A program written in a semi-formal fashion such as Pidgin-Pascal, and

(v) A formal program capable of being executed by a computer.

We now consider the input necessary in order to execute a given algorithm.

9.2 Input

Algorithm input involves the specification and communication of the information necessary for the implementation of the algorithm on a given problem instance. In the case of graph theoretic algorithms, this involves communicating the structure of a number of graphs, along with certain parameters (such as their edge weights), and also bounds.

There are a number of options for communicating the structure of a graph or digraph. There is no one option which is uniformly superior in utility — the choice depends upon the structure of the graphs to be communicated, the problem to be resolved, the algorithm to be used, and how the algorithm is to be implemented (i.e. by hand or by what type of computer and what computer language). Some of the most common options are:

i) The Adjacency Matrix (See Section 6.1)

Because A is symmetric for any graph, it is sometimes worth storing only its upper or its lower triangle. If the graph to be inputted has n vertices, this reduces the number of entries to be stored from n^2 to $n(n-1)/2$.

Multigraphs (with parallel edges) can be stored by redefining a_{ij} (the $i-j$th entry of A) to be equal to the number of edges incident with both vertices i and j.

ii) The Incidence Matrix (See Section 6.2)

It is often worthwhile to store digraphs with relatively few edges by means of their incidence matrices. This is because the number of entries that must be stored is ne, where e is the number of arcs in the digraph.

iii) Edge Listing

In this input form, each edge of a pseudograph or digraph is specified as an unordered pair of its incident vertices. Once again, this vehicle may be worthwhile if there are relatively few edges.

iv) Linear Arrays

If a weighted pseudograph or digraph is to be inputted, it may be efficient to specify three vectors:

$$U = (u_1, u_2, \ldots, u_n),$$
$$V = (v_1, v_2, \ldots, v_n), \quad \text{and}$$
$$W = (w_1, w_2, \ldots, w_n)$$

in which the i-th edge (or arc) is $u_i v_i$ with weight w_i. As an example, the weighted graph in Figure 9.1 has, as one of its many linear array representations:

$$U = (1, 1, 1, 2, 3),$$
$$V = (2, 3, 4, 3, 4), \quad \text{and}$$
$$W = (10, 29, 32, 20, 1).$$

v) Adjacency Lists

When depth-first search (to be discussed later) is used or paths are to be found in a digraph, it is often useful to represent the digraph by a sequence of vectors, one for each vertex i, in which the i-th vector contains as its first entry v_i, and as later entries, the vertices for which there exist incident

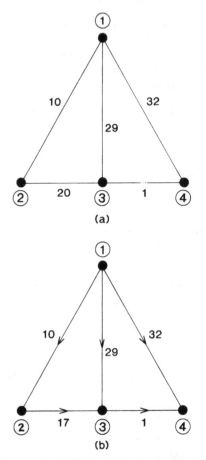

Figure 9.1 (a) A weighted graph and
(b) a weighted digraph

arcs adjacent away from v_i. As an example, the weighted digraph has as its adjacency list:

$$(1, 2, 3, 4), (2, 3), (3, 4), (4).$$

Adjacency lists can be used to represent pseudographs (with multiple arcs and loops). They are less efficient for graphs, as every edge is represented twice.

Quite often, especially when more than one analysis is to be made of a graph, it is worthwhile to convert one input representation into another. As each of the five representations presented above completely specify a given graph, it is a simple matter to make such a conversion. We now consider the efficiency of an algorithm.

9.3* Complexity

We now study a concept related to the efficiency of an algorithm, called *complexity* (occasionally called *time complexity*). An *elementary computational* step is an operation performed on two numbers which is one of: addition, subtraction, multiplication, division, or size comparison.

The *complexity* of an algorithm is proportional to the maximum number of elementary computational steps required to transform the input for the algorithm into the result of the implementation of the algorithm on that input, taken over all possible problem instances.

Complexity is usually a function C, of a quantity commonly called *problem size*, denoted by s, which is the number of bits of input data. In graph theoretic algorithms, problem size is a function of either n or e, or both, where n and e are: the number of vertices and the number of edges respectively of the inputted graph(s).

Here we have defined *complexity* in terms of the maximum number of steps required by an algorithm, taken over all possible problem instances of a given problem size. This is known as *worst-case complexity*, as opposed to *expected time complexity* (*average time complexity*). It may be that the complexities of two given algorithms a_1 and a_2, for a given problem, differ. As an example, suppose that $C_1(s) = 6s + 12$, is the complexity function for a_1 and $C_2(s) = s^2 + s + 6$ is the complexity for a_2 where s is a nonnegative integer.

Then for $s < 6$, a_2 is faster (requires fewer steps) and, for $s > 6$, a_1 is faster.

In fact, no matter what coefficients C_1 and C_2 possess, there exists a value of s, say s_0, with the property that a_1 is faster than a_2 for all $s > s_0$. In this case, the complexity of a_1 is said to be of a *lower order* than that of a_2.

Definitions

If C_1 and C_2 are the complexity functions of two given algorithms a_1 and a_2 respectively, it is said that C_1 has *order no greater than* that of C_2 if there exist numbers s_0 and K such that

$$C_1(s) \leq KC_2(s), \qquad \text{for all } s > s_0.$$

This is expressed as $C_1 = O(C_2)$.

If $C_1 = O(C_2)$ and $C_2 = O(C_1)$, then C_1 and C_2 are said to be *of the same order*.

As an example, $4n^4 + 6n^3 + n^2 + n + 3 = O(n^4)$. The order of an algorithm with complexity function $C(s)$, describes the behaviour of $C(s)$ as $s \to +\infty$.

If $C_1 = O(C_2)$ then C_1 is said to be an $O(C_2)$ algorithm.

The reader may have been wondering where this development is leading to. It turns out that complexity is an important concept for many graph theorists, especially those interested in graph theory applications. This is because it is often important to have a sense of the limitations of an algorithm in terms of its ability to solve a problem instance of a given size in a reasonable amount of computational time. This arises from the fact that certain algorithms cannot guarantee to solve, in a practical sense, general problem instances greater than a rather small size. These algorithms may be essentially no better that the approach of completely enumerating all solution possibilities. With the fastest computers available, or even envisaged, the required number of computational steps (and hence time) is so large that the algorithms are capable of solving only trivial problem instances.

This means that it is useful to be able to recognise which algorithms, because of their efficiency, are worth employing, and which algorithms are so inefficient that they are of, at most, theoretical interest.

Definition

An algorithm a with
$$C_a = O(C),$$

is said to be of *polynomial, exponential,* or *factorial order* if C is a polynomial, exponential, or factorial function respectively. Polynomial algorithms are sometimes called *good* algorithms because, as problem size increases, the number of steps required by a polynomial-order algorithm, (as opposed to exponential-or factorial-order algorithms) grows relatively slowly. Indeed, no matter what coefficients or order their complexity functions have, there always exists a problem size say, s_0, such that any polynomial order algorithm will require fewer steps for all problems of size at least s_0, than any exponential or factorial order algorithms. However it is important to note that here we are considering *worst-case performance*. There exist certain problems (such as the linear programming problem) for which the *average case performance* of a polynomial-order algorithm is often worse than that of certain exponential-order algorithms for relatively large problem size. One of the reasons for this is that the definition of a good algorithm takes no account of the fact that the polynomial giving rise to the order of a polynomial algorithm may have very large coefficients.

Definition

A problem for which it is conjectured that no polynomial algorithm exists is said to be *intractable*. Despite the fact that there is an $O(n^2)$ algorithm for the *shortest path problem* (see Chapters 10 and 12), the *longest path problem* is intractable when nonnegative edge weights are assumed.

It is useful to state a *decision problem* version of intractable optimization problems. A decision problem is one whose solution is a simple answer of *yes* or *no*. Any optimization problem can be turned into a decision problem by the use of a bound. As an illustration, consider the intractable problem just mentioned.

The Longest Path Optimization Problem

Given a graph $G = (V, E)$, nonnegative edge weights, and given $u, v \in V$ where $u \neq v$, find the path P^*, between u and v of maximum weight, as a sum of the weights of the edges in P^*.

The optimization problem can be converted into a decision problem by introducing a bound as follows:

The Longest Path Decision Problem

Given an integer B, a graph $G = (V, E)$, nonnegative edge weights, given $u, v \in V$ where $u \neq v$, does there exist a path between u and v of weight at least B?

The optimization version of the problem is solved by repeatedly solving the decision version of the problem, by incrementally decreasing the value of B. When the answer changes from *no* to *yes*, the solution is at hand.

This problem belongs to a class of problems called *nondeterministic polynomial time complete*, or *NP-complete*. All the problems in these classes are believed to be intractable. Many of the best algorithms (in terms of time complexity) for the NP-complete problems comprise a number of polynomial-order sub-algorithms. However they are inefficient because the subroutine must be used a number of times, where the number is of exponential order. As an example, consider the following intractable problem, which is discussed further in Section 14.2.2.

Maximum Weight Planar Subgraph

Given a weighted graph $G = (V, E)$, find the planar subgraph of G with maximum total edge weight.

There exists an algorithm due to Hopcroft and Tarjan (1974) for detecting whether or not a graph G is planar, of order $O(n)$, where n is the number of vertices of G. The basic ideas of this algorithm are outlined in Section 9.5. However the best algorithms when implemented for general instances of this problem require that this subroutine be used a number of times, where the number is of exponential order.

It can be shown that the solution to any given NP-complete problem instance can be transformed into that for any other NP-complete problem in a number of steps which is of polynomial order. This means that if an algorithm of polynomial order can be found for any NP-complete problem, then all the NP-complete problems can be solved by polynomial algorithms. There is no evidence to suggest that such an algorithm exists.

Later in this section we shall provide a formal definition of the concept of an *algorithm*, based on what is known as a *Turing machine*. Our study of algorithms is greatly aided by the use of this machine as a model of computation because of its simplicity. We also take advantage of the fact that the addition or multiplication of two polynomials results in a further polynomial, which is especially useful in proving that certain algorithms are of polynomial order. We now place the notions introduced so far in this section on a more rigorous footing.

A *Turing Machine TM*, is a computing machine which performs its calculations on an infinite tape which is divided lengthwise into cells. The number of symbols in any cell is always exactly one. Input to the machine comprises a tape which is blank except for a finite number of neighbouring filled cells. Calculations are performed by a *tape head* which has a finite number of internal states. They are carried out in a cyclic fashion by the tape head reading the cell C, opposite it. The tape head then:

   (i) Overwrites the contents of the cell with a new symbol (which may be identical to the previous symbol),

   (ii) Enters a new state (which may be identical to the previous state),

   (iii) Moves the tape so that the neighbouring cell to the immediate left or immediate right of C is opposite the tapehead, and

   (iv) Begins a new cycle by reading the new cell.

What is overwritten in (i), what state is entered in (ii), and the direction of movement in (iii) depends upon:

   (a) The current state of the tape head,

   (b) The symbol read, and

(c) The value of a variable m, specifying the direction of movement which is calculated from (a) and (b).

We assume that the total number of symbols that the cells can contain is finite. We also assume temporarily that from any pair of (i) a present tape head state, and (ii) a symbol just read, TM can determine exactly one new tape head state symbol to be written, and one direction of movement. That is, TM is *deterministic*. It is denoted by *DTM*.

An initial state and cell to be read must be specified for calculations to begin. A set of *final* or *halting* states are specified. When TM reaches a final state, calculations are terminated. The output of the calculations is provided by either the symbols printed on the tape or the final state, or both. A typical final state is either *yes* or *no*.

We now confine ourselves for the rest of this section to decision problems. The class of decision problems for which there exists a polynomial DTM algorithm is called *polynomial* and is denoted by *P*.

Earlier in this chapter, we stated that there is no reported efficient algorithm for the longest path decision problem, i.e the problem is not known to be polynomial. Consider however, a given problem instance in which a particular arc set was produced for which it was claimed that it was a path whose length was greater than the given bound, i.e. the answer to the decision problem for this particular path is *yes*. It is straightforward to devise a *polynomial* verification algorithm to check whether or not the arc set does indeed correspond to a path (between the required vertices) of sufficient length. If it is possible to implement this verification algorithm *simultaneously* to all candidate paths then the decision problem could be solved in polynomial time. Many of the important decision problem versions of the problems discussed in Part II of this book are of this nature. That is, there is no efficient algorithm reported for any of them, but they are *polynomial-time verifiable*.

This class of problems is termed *nondeterministic polynomial*. Thus a problem is *nondeterministic polynomial* if we can provide a polynomial DTM algorithm which will terminate when applied to any *guess* (e.g a given edge set for the longest path problem) for any decision problem instance (e.g. a given weighted graph and a given bound).

We now build up the concept of a two-phase nondeterministic algorithm for a TM. A problem instance is assumed written in the cells according to the nondeterministical input convention recorded earlier. The first phase constructs a guess and records it in the cells according to the same input convention. The construction is nondeterministic in the sense that, when choices have to be made in the process, TM makes a choice arbitrarily. The second phase verifies deterministically whether the answer *yes* or *no* should be assigned to the guess for the inputted problem instance.

A TM which operates such an algorithm is termed a *Nondeterministic Turing Machine, NDTM*. Consider a NDTM which is applied to a decision problem DP. Consider a problem instance I, of DP with size s. Then NDTM is said to be *a polynomial time NDTM* if, for all instances of I which result in the answer *yes*, there is a guess that causes the verification algorithm of NDTM to terminate in a number steps which is a polynomial function of s. The class of all problems which can be solved by a polynomial time NDTM are called *nondeterministically polynomially bounded* and is denoted by NP. Clearly $P \subseteq NP$. It seems likely that P is a proper subset of NP.

We turn now to the concept of *polynomial transformation*. Consider two decision problems, DP_1 and DP_2.

Definition

There exists a *polynomial transformation* from DP_1 to DP_2, denoted by $DP_2 \propto DP_1$, if:

(i) There exists a function f, transforming any instance I, of DP_1, to an instance of DP_2 such that the answer corresponding to I with respect to DP_2 is *yes* if and only if the answer corresponding to $f(I)$ with respect to DP_2 is *yes*, and

(ii) There exists a polynomial algorithm to implement f.

As a result of this definition we have

Theorem 9.1

If DP_1 and DP_2 are two decision problems and there exists a polynomial algorithm for DP_1, and $DP_2 \propto DP_1$, then there is a polynomial algorithm for DP_2.

Proof

Suppose instances I_1 and I_2 of DP_1 and DP_2 respectively, have sizes s_1 and s_2 respectively. As $DP_2 \propto DP_1$, there exists a function f, along with a polynomial algorithm for implementing it, such that $I_1 = f(I_2)$. Also the functions bounding the computation of I_1 ($= f(I_2)$) and of I_2 are necessarily polynomial. We denote them by $P_1(s_2)$ and $P_2(s_2)$ respectively. We have $s \leq P_2(s_2)$. Thus the calculation time for DP_2 (i.e. the time for the transformation to DP_1 plus the time for the calculation of DP_1) is bounded by $P_2(s_2) + P_1(P_2(s_2))$. As this is a simple function of polynomials, it is itself polynomial in s_2. ■

Corollary 9.1

"$\propto$" is transitive. That is, if $DP_1 \propto DP_2$ and $DP_2 \propto DP_3$, then $DP_1 \propto DP_3$.

Definition

A decision problem DP, is said to be *NP-hard* if for every problem $DP_1 \in NP$, $DP_1 \propto DP$; if in addition, $DP \in NP$ then DP is said to be *NP-complete*.

This means that if there exists an NP-hard problem which is polynomial, then all NP-hard problems are polynomial, i.e $P = NP$. Also if a decision problem DP, is NP-hard and there exists another decision problem DP_1 which is in NP, and for which $DP \propto DP_1$, then DP_1 is NP-hard.

In this sense the NP-hard problems are the hardest in NP and form an equivalence class.

We now present a theorem stating that a known decision problem is NP-hard. By using this theorem (or similar ones stating that other decision problems are NP-hard) and the fact that the relation of polynomial transformation is transitive, many problems can be shown to be NP-hard. We use the theorem to prove that another decision problem is NP-hard as an illustration later. First we state the decision problem.

Exact 3 Cover (X3C)

$$\text{Input} \quad \mathcal{F} = \{F_1, F_2, \ldots, F_n\}, \quad \text{where } |F_i| = 3 \quad \text{and}$$
$$F_i \subseteq \{1, 2, \ldots, 3m\} = I, \quad i = 1, 2, \ldots, n.$$

Question: Does $\mathcal{F}$ contain 3-sets $F_{i(1)}, F_{i(2)}, \ldots, F_{i(3)}$, whose union is I_{3m}? Note that if $\mathcal{F}$ does contain m such 3-sets, then they must be disjoint.

Theorem 9.2

X3C is NP-hard.

Proof

The reader is referred to the proof given by Garey and Johnson (1979) .

We now introduce another decision problem which we shall show is NP-hard by using Theorem 9.2. This problem, called the *Steiner Problem in Phylogeny*, is further discussed in Section 15.3.

Let $G = (V, E)$ be a graph with weight function $w : E \to \Re$. For any $X \subseteq V$, we define a *minimum spanning tree* $T(X)$, *for* X to be a tree of G with vertex set containing X such that the sum of the edge weights of $T(X)$ is a minimum. Also, we define a *Steiner minimal tree* $S(X)$, *for* X, to be a tree having the *minimum possible* weight over all trees in G which contain X in their vertex sets.

It is well known that for arbitrary weighted graphs, finding a Steiner minimal tree (SMT) is, in general, NP-hard. More recently, it has been shown that for graphs whose edge weights come from certain metric structures, such as the Euclidean plane or the L_1 plane, finding SMT's is also NP-hard. (Garey et al. (1977), Garey and Johnson (1977, 1979).)

The problem we are considering, i.e. the Steiner Problem in Phylogeny has the following formalization. For a fixed alphabet A, let d denote the Hamming distance on A^N (the Cartesian product $A \times A \times, \ldots, \times A$, where there are N instances of A in the product), i.e., $d((a_1, \ldots, a_n), (a'_1, \ldots, a'_n))$ is equal to be the number of indices i, such that $a_i \neq a'_i$.

In the metric space $(A^N d)$, the *Steiner problem in phylogeny (SPP)* is:

$\quad$ (**SPP**): **Given a set $X \subseteq A^N$, find a Steiner minimal tree** $S(X)$, **for** X.

We shall show that, even when A consists of just two elements, the SPP for A^N is NP-hard. (Strictly speaking, we should really be considering the decision problem of deciding whether X has a Steiner tree with length at most some prespecified bound B.) Let $A = \{0, 1\}$. For a fixed positive integer N, denote A^N by Q_N. The graph $G = G(Q_N, D)$ is just the 1-skeleton of the N-cube. To show that the Steiner problem for Q_N is NP-hard (which we shall denote by SPQ), we shall reduce the NP-hard problem, Exact 3-Cover (X3C) to SPQ.

We now give the details for the construction of the desired corresponding instance of SPQ. To begin with we set $N = 4m(n + 3m + 1)$. A vertex $q = (q_1, \ldots, q_N) \in Q_N$ can be thought of as consisting of $n + 3m + 1$ blocks, each of length $4m$:

$$q = (X_0, X_1, \ldots, X_{3m}; Y_1, \ldots, Y_n). \tag{9.1}$$

To each integer i, $0 \leq i \leq 3m$, define a point x_i by taking in (9.1)

$$X = (1, 1, \ldots, 1) \equiv \overline{1} \quad \text{(with 4m unit entries)}.$$

and all other X_j and all Y_k to be

$$(0, 0, \ldots, 0) \equiv \overline{0} \quad \text{(with 4m zero entries)}.$$

Similarly, for $1 \leq j \leq n$, define s_j to have $X_i = \bar{0}$ for all i, $Y_j = \bar{1}$ and $Y_k = \bar{0}$ for all $k \neq j$. Intuitively, for $i \neq 0$, x_i will correspond to the integer i and Y_j will correspond to the 3-set F_j.

Next, if $i \in F_j$, we define a sequence of points $x_{i,j}(k)$, $0 \leq k \leq 8m - 1$, as follows:

$$x_{i,j}(k) = (X_0, \ldots, X_i(k), \ldots, X_{3m}; Y_1, \ldots, Y_j(k), \ldots, Y_n),$$

where $X = \bar{0}$, $u \neq i$, $Y_v = \bar{0}$, $v \neq j$ and

$$X_i = 1, \overline{Y}_j^{(k)} = (0, 0, \ldots, 0, 1, 1, \ldots 1), \quad 0 \leq k \leq 4m,$$

where the first $(4m - k)$ entries are zero entries and the remaining k entries are unit entries.

$$X_i^{(k)} = (0, 0, \ldots, 0, 1, 1, \ldots, 1), \quad 4m < k \leq 8m - 1, Y_j = \bar{1},$$

where the first $(k - 4m)$ entries are zero and the remaining $(8m - k)$ entries are unit entries.

Also, define $x_{0,j}(k)$ as above for all j, k, where $1 \leq j \leq n$, $0 \leq k \leq 8m - 1$.

Note that $x_{i,j}(0) = x_i$ for $1 \leq i \leq 3m$. Observe (for future reference) that the $x_{i,j}(k)$ form *chains* from x_i to s_j, where consecutive points on the chain have distance 1.

The set $X = X(\mathcal{F})$ comprises the $8m(3m + 1)n$ points $\{x_{i,j}(k) : 0 \leq i \leq 3m, 1 \leq j \leq n, 0 \leq k \leq 8m - 1\}$. We point out that X has a *spanning tree* with maximum edge length 2. This implies (see Garey et al. (1977)) that any edge in an SMT for X has length at most 2. Define

$$L_0 = 4n(8m - 1) + 4m,$$

and let $L_s(X)$ denote the length of an SMT for X.

Theorem 9.3

If $\mathcal{F}$ has an X3C then $L_s(X) \leq L_0$.

Proof

Let $F_{j(1)}, \ldots, F_{j(m)}$ be an exact 3-cover of I_{3m}. For $1 \leq k \leq m$, adjoin to X the Steiner points $s_k = (X_0, \ldots, X_{3m}, Y_1, \ldots Y_n)$ with $X = \bar{0}$, $1 \leq i \leq 3m$, $Y_{j_k} = \bar{1}$, $Y_j = \bar{0}$, $j \neq j_{j_k}$. An easy calculation shows that $X_+ = X \cup \{s_k : 1 \leq k \leq m\}$ has a spanning tree of length L_0; just form the spanning tree consisting of all length 1 edges for X_+. ∎

Let us call an SMT for X, *greedy* if it uses all the length 1 edges between points in X.

Theorem 9.4

X has a *greedy* SMT with length $L_s(X)$.

Proof

Suppose T is an SMT for X with length strictly less than any greedy SMT. Thus, some edge e, of length 1, does not occur in T. Adjoin e to T, thereby forming a cycle C. Some edge e', in C must be incident to a Steiner vertex s, of T (since the length 1 edges in X do not form any cycles). Form the SMT, T', by deleting the edge e'. Since

$$\text{length}(e') \geq 1 = \text{length}(e).$$

then

$$\text{length}(T') \leq \text{length}(T).$$

However T is, by hypothesis, an SMT. So that in fact

$$\text{length}(L') = \text{length}(T) = L_s(X).$$

Note that T' contains one more length 1 edge between points of X than T has. The result now follows by induction. ■

Theorem 9.5

If $L_s(X) \leq L_o$ then $\mathcal{F}$ has an X3C.

Proof

By Theorem 9.4, we may assume that X has a greedy SMT, T', with length$(T') \leq L_0$. As usual, we can assume, without loss of generality, that every Steiner vertex of T' has degree at least 3. For $0 \leq i \leq 3m$, define the subtree T_i of T' to be the tree induced by the points $x_{ij}(k)$, $1 \leq j \leq n$, $0 \leq k \leq 8m - 1$. By construction, all edges of T_i have length 1. We can think of T' as being formed by connecting the T_i together with a set E of edges, each of which is incident to some Steiner vertex.

Observe that since

$$\sum \text{length}(T_i) = 4n(8m - 1) \tag{9.2}$$

then

$$\sum_{e \in E} \text{length}(e) \leq 4m \tag{9.3}$$

A key fact to be noted is that for any $i_1 < i_2 < i_3 < i_4 < i_5$ and any $q \in Q_N$

$$\sum_{k=1}^{5} d(q, T_{i(k)}) > 4m, \qquad (9.4)$$

where $d(q, T_i)$ denotes the minimum distance from q to a point of T_i. (This is the reason that subblocks of length $4m$ are used in the definition of the $x_{i,j}(k)$).

A consequence of this observation is that no component of E can have more than two Steiner vertices. Otherwise, some Steiner point of T' would be connected by paths in E to (at least) five different T_i's which, by (9.3), would force

$$\text{length}(T') > 4n(8m - 1) + 4m = L_0.$$

Thus there are at most four types of connected components E_k, which can be formed by edges from E (also called *full Steiner subtrees*; see Garey et al. (1977)). They are illustrated in Figure 9.2.

In Table 9.1 we list for each case, a lower bound on the length of E_k, the *decrease* $\Delta(E_k)$, in the number of components due to E_k, and ρ, a lower bound on the ratio $\text{length}(E_k)/\Delta(E_k)$. Note that for all E, $\text{length}(E) \leq 4m$ and $\Delta(E) = 3m$; thus $\rho(E) \leq 4/3$.

Case	Lower Bound on Length(E_k)	$\Delta(E_k)$	ρ
(i)	2	1	2
(ii)	3	2	3/2
(iii)	4	3	4/3
(iv)	5	3	5/3

Table 9.1 Analysis of the four cases.

Therefore, the only possibility is that case (iii) holds (with equality) in *all* cases. In other words, T' must have m Steiner vertices, each of degree 4, with all connecting edges of length 1. However, this is only possible if these Steiner vertices are m of the s_j's, i.e., with $X_j = \bar{0}$ for all i, $Y_{jk} = \bar{1}$, $Y_j = \bar{0}$, $j \neq j_k$. Consequently, the F_j's form an exact 3-cover of I_{3m}. ∎

The preceding three theorems have, as immediate consequences, the following results.

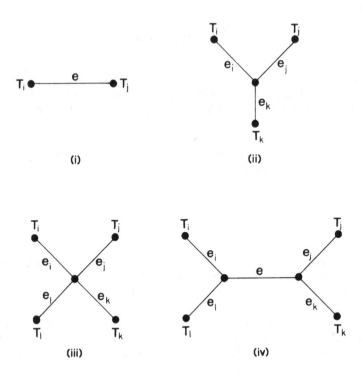

Figure 9.2 Possible components of E

Theorem 9.6

The Steiner problem for the N-cube Q_N, is NP- hard.

Corollary 9.2

The Steiner problem in phylogeny is NP-hard.

9.4 Output

There would be little point to an algorithm if it did not produce some output. Whereas the input to graph theoretic algorithms is invariably a collection of graphs (possibly one) and maybe some parameters and a bound, the output is highly dependent upon the type of problem. We have introduced decision problems in which the output will be either *yes* or *no*. In other instances we may require, as output, some parameter of a graph, such as its rank, thickness, chromatic number, or shortest spanning tree weight. In more complicated instances we may require not just a simple response to a question, but also a subgraph of the inputted graph which justifies the response. As an example we may wish to discover, not only whether a graph has a perfect matching, but also to find a perfect matching in the graph , if it exists.

We now discuss algorithms which analyse, rather than optimize a given graph.

9.5 Graph analysis algorithms

Solving practical problems with graph theoretic techniques often involves analysing an underlying graph and testing it for a given property. Because graphs arising from graph theoretic models of practical problems are usually relatively complex, it is important to use efficient analysis algorithms suitable for implementation on a digital computer. To this end we now discuss graph analysis algorithms for the following topics:

(i)   Tree search,
(ii)   Connectivity,
(iii) Planarity,
(iv)   Spanning trees,
(v)   Isomorphism, and
(vi)   Fundamental cycles.

9.5.1 Tree search methods

In the analysis of a graph G, which models a real problem, it is often necessary to examine, at least implicitly, every edge of G. The two obvious ways of examining all the edges of a graph G, are:

(i) At vertex v of G, examine all the edges incident with v and then pass to another vertex of G which is adjacent to v, or

(ii) At vertex v of G, examine a single edge incident with v and then pass to the other vertex in G which is incident with this edge.

The first of these ways is called *breadth-first search (BFS)* and the second is called *depth-first search (DFS)* or *backtracking*.

Breadth-first search

The BFS approach is used in the algorithms for identifying: connectivity (Section 9.5.2), fundamental cycles (Section 9.5.6), and shortest paths (Section 12.5.2). In general, the edge examination process can be represented by a *search tree* T, with each vertex in the tree representing a vertex of the graph G, to be analysed. An initial vertex, say v, of G is chosen (at random if there is no better choice suspected). It represents the initial vertex in T, representing the initial, zeroth level of T. Each of its incident edges are examined in turn and are represented by edges incident with v in T. The new vertices incident with these edges represent the first level of T. Each of the unexamined edges, incident with vertices adjacent to v are examined if this is fruitful. Thus a series of walks are identified in G, each represented by a path in T, starting at v. Search paths in T are terminated when either: (i) there are no further incident edges to be examined, or (ii) further examination would be fruitless. A BFS tree for the graph of Figure 9.3(a) is shown in Figure 9.3(b).

Depth-first search

The DFS approach is used in the algorithms for identifying: planarity (Section 9.5.3), spanning trees (Section 9.5.4), isomorphism (Section 9.5.5), fundamental cycles (Section 9.5.6), and in many other analyses. A search tree T, can once again be used to represent the edge examination process. In BFS, each edge incident with a selected vertex v, is examined before selecting a new adjacent vertex. By contrast, in DFS a new adjacent vertex is selected, which is incident with the first edge incident with v. In other words, in BFS we stay at v as long as possible, examining all of its incident edges, before selecting a new vertex adjacent to v. In DFS we leave v as quickly as possible, examining only one of its incident edges and replacing v by a new vertex, which is adjacent to v. This may leave edges incident with v which are unexamined. These edges may have to be examined later, if the analysis is not resolved in the meantime, or if it is necessary to examine every edge, come what may.

We now describe a DFS algorithm for the examination of every edge of a connected graph $G = (V, E)$, with n vertices. During the DFS process let:

  (i) $N(v) = i$ imply that vertex v ($\in V$) was the ith vertex to be visited.

  (ii) P, F be a partition of E (to be constructed). (P stands for *palm* and F for *frond*.)

If G has n vertices, then DFS will terminate having labelled the vertices with,

$$\{N(v_1), N(v_2), \ldots, N(v_n)\} = \{1, 2, \ldots, n\},$$

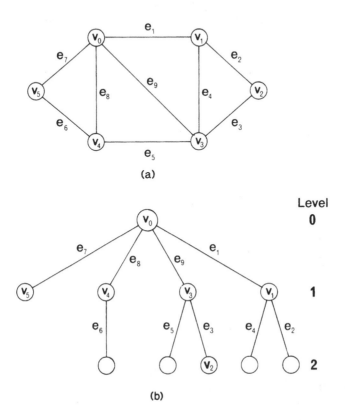

Figure 9.3 (a) A graph and (b) its BFS tree

in a one-to-one fashion. Every edge of E will have been given an *orientation* and thus G will have been transformed into an oriented graph, say digraph $D = (V, A)$. The set P, when edge orientation is taken into account, constitutes a spanning arborescence, called a *palm tree*. F comprises the edges of E which are not in P. If $uv \in A$ then $N(u) > N(v)$.

The contribution of DFS is not to analyse a given graph directly for any particular property, but rather to provide a scientific vertex labelling and a spanning arborescence. These contributions make possible the construction of very efficient algorithms for certain graph analysis problems.

DFS algorithm statement

Let $G = (V, E)$ be a connected graph, v_0 be a vertex of V at which the process is to begin, and v be the current vertex being examined.

STEP

(1) Set v to become $\quad v_0,$
  $\qquad\qquad P \qquad\qquad\quad \phi,$
  $\qquad\qquad F \qquad\qquad\quad \phi,$ and
  $\qquad\qquad i \qquad\qquad\quad 0.$

(2) Set i to become $\quad i + 1,$ and
  $\qquad\qquad v \qquad\qquad\quad v_i.$

(3) Attempt to identify an unexamined edge of E incident with v.
  (i) If no such edge exists go to step (5), otherwise
  (ii) Choose any such edge, say vu, and examine it. Orient this edge as the arc (v, u). Go to step (4).

(4) Check whether $N(u)$ is defined.
  (i) If $N(u)$ is undefined,
  Set P to become $\quad P \cup \{(v, u)\},$ and
  $\qquad v$ to become $\quad u.$
  Go to step (2).

  (ii) If $N(u)$ is defined,
  Set F to become $\quad F \cup \{(v, u)\}.$
  Go to step (3).

(5) Check whether there exists an examined arc, say $(t, v) \in P$.
  (i) If such an arc (t, v), exists,
  Set v to become $\quad t$
  $\qquad\quad$ Go to step (2).
  (ii) If such an arc does not exist, terminate.

The vertex numbering assigned to the vertices of a graph by the DFS algorithm is the opposite of a *logical numbering,* explained in Section 7.6.2.

We now illustrate the DFS algorithm by applying it to the graph shown in Figure 9.3(a), with initial vertex v_0. The resulting oriented graph is shown in Figure 9.4(a) with the palm tree shown in solid lines and the fronds shown in broken lines. The DFS tree is shown in Figure 9.4(b).

9.5.2 Connectivity

In many applications of graph theory to practical problems, one often first wishes to establish whether or not a given graph is connected and, if it is not, to identify all of its components. This process is often useful as a subroutine in other graph theoretic algorithms and is far from trivial for

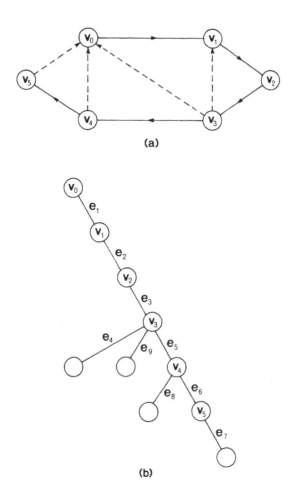

Figure 9.4 (a) An oriented graph and (b) its associated DFS tree

complex graph models for which no diagram is available. One approach is to manipulate the adjacency matrix A, of the graph in question in order to see if its rows or columns can be permuted so as to admit a block diagonal version of A. (See observation A(iv) in Section 6.1.)

Another approach involves *coalescing* a vertex v, with all vertices adjacent to it in the given graph, by removing all edges between them and replacing the vertices by v. The coalescement process is illustrated in Figure 9.5(a) with vertex v_0.

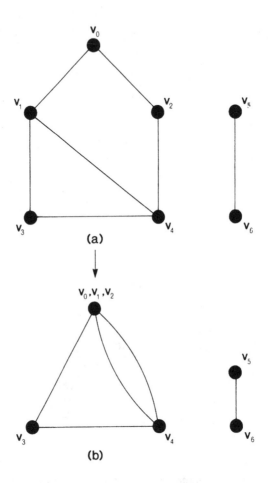

Figure 9.5 The coalescement process

The coalescement process is repeated in the multigraph produced until further coalescement is impossible. The result is an edge-less graph whose vertices represent the components of the original graph.

It appears to be efficient to choose to be coalesced next, the vertex with maximum degree in the current component being identified. The strategy can be achieved by manipulating the adjacency matrix A, of the given graph as follows. If the vertex v_i is to be fused with vertices $v_j, v_k, \ldots$, then add the jth, kth,..., rows (columns) to the ith row (column) of A and discard the jth, kth,..., rows and columns of A.

When no further coalescement is possible, the positive entries in the final matrix correspond to the components and equal twice the number of edges in each component.

DFS Connectivity Algorithm Statement

Let G be the graph to be analysed,

         G' be the graph created from G by coalescement at each iteration,

         v be the current vertex being examined, and

         $N(v_i)$ be the component number to which vertex i belongs.

Step

(1)   Set i to become 0,

             G' to become G.

(2)   Set v to become the unlabelled vertex with the highest degree.

(3)   Check whether $d(v) = 0$ in G'.

      (i)   If $d(v) = 0$ go to step (6).

      (ii)   If $d(v) > 0$,

      Set i to become $i + 1$, and

            $N(v)$ to become i.

      Go to step (4).

(4)   Set $N(v_j) = i$, $\forall v_j \in \Gamma(v)$. In G' coalesce v with $\Gamma(v)$ in G'.

(5)   Check whether $d(v) = 0$ in G'.

      If $d(v) = 0$, go to step (2).

      If $d(v) > 0$, go to step (4).

(6)   Label each unlabelled vertex in G with $i + 1, i + 2, i + 3, \ldots$

      Go to step (7).

(7)   Terminate. The vertices labelled i comprise the ith component of G.

When this algorithm is applied to the graph in Figure 9.5(a) it results in vertices: v_0, v_1, v_2, v_3, and v_4, being identified as component 1 and vertices v_5 and v_6 as component 2. This process corresponds to the following operations being performed on the adjacency matrix when its rows and columns are added together.

	0	1	2	3	4	5	6
	0	1	1	0	0	0	0
1	1	0	0	1	1	0	0
2	1	0	0	0	1	0	0
3	0	1	0	0	1	0	0
4	0	1	1	1	0	0	0
5	0	0	0	0	0	0	1
6	0	0	0	0	0	1	0

	1	2	5	6
1	8	2	0	0
2	2	0	0	0
5	0	0	0	1
6	0	0	1	0

	1	5	6
1	12	0	0
5	0	0	1
6	0	1	0

	1	5
1	12	0
5	0	2

9.5.3 Planarity

The characterizations of planarity given in Theorems 5.8 and 5.11, although elegant, do not provide a practical basis upon which to devise an efficient graph planarity detection algorithm for general graphs. However a DFS algorithm, with complexity $O(n)$, where n is the number of vertices in the graph to be tested, has been presented by Hopcroft and Tarjan (1974) . We give only an outline of the algorithm. The reader is referred to the previously-mentioned paper for the full details.

Recall the DFS algorithm of Section 9.5.1. In applying it to analyse a given graph $G = (V, E)$ for planarity, a palm tree will be identified. The first frond identified is added to this tree to create a cycle C, in G. The removal of C from G gives rise to a number of fragments of G which are to be tested for planarity.

The algorithm relies upon a subroutine which records the paths in the DFS decision tree which each comprise a sequence of palm edges followed by a single frond. The logic of the algorithm can be followed by imagining an attempt to build up a plane drawing of the graph as the DFS unfolds. First C is embedded in the plane. Each DFS path joins two vertices in C by a number (possibly zero) of new vertices. Sometimes there is only one feasible way to draw a new DFS path P and preserve planarity.

The addition of a path P', to C will give rise to two cycles in the plane. An additional path P', can be added if it is part of exactly one existing cycle. If this is not the case, there are two regions in which P' can be drawn. Because of this ambiguity, P' is placed on a stack to be drawn later. The

DFS algorithm then identifies another path P'', which begins at one of the vertices of P'.

P'' may join vertices on two different paths in the current diagram. If these paths are not on the boundary of a common face then the diagram cannot be completed in a planar fashion and thus G is nonplanar. If not, these paths are drawn.

If P'' joins vertices on the same drawn path, then P'' must also be placed on the stack, as it is not clear as to where to draw it.

If P'' joins a vertex on a drawn path to a vertex not yet present in the drawing, it may make definite where P' must be drawn. In this case P' and P'' are both drawn.

The algorithm continues in this way, and at each iteration selects a new path which it either:

(i) Adds to the plane diagram, or

(ii) Places on the top of a push-down stack to be drawn later, or

(iii) Rejects as impossible to add, resulting in a pronouncement of non-planarity.

Because a planar graph can possess at most $(3n - 6)$ edges (Corollary 5.2) only $(3n - 6)$ edges need be examined. The algorithm is $O(n)$ because it can be shown that the number of elementary computational steps required to examine each edge is no more than a constant.

9.5.4 Spanning Trees

It is often of interest to be able to identify a spanning tree in a connected graph and a spanning forest in an arbitrary graph. Examples of this will be given in Section 13.1 on electrical networks. Clearly, if a spanning tree has been identified in a given graph, then the graph has been shown to be connected. However, the connectivity algorithm given in Section 9.5.2 is, in general, a more efficient test for connectivity than the spanning tree algorithm to be explained in this section. In Section 12.5.1, two algorithms are presented for the detection of a minimal spanning tree in a weighted connected graph.

We now describe a DFS algorithm for identifying a spanning forest in a given graph. It is assumed that the graph G, is described by means of n adjacency lists (see Section 9.2) of vertex labels, where n is the number of vertices of G. The ith list comprises the vertices which are adjacent to v_i, for each vertex v_i in G.

The DFS (see Section 9.5.1) approach is used to create an order in which the edges of G are examined. As each edge is examined, it is accepted as part of a spanning forest being constructed, unless it creates a cycle with the edges accepted. In this latter case the edge is rejected and the next edge is examined. The spanning arborescence created by the DFS represents the spanning tree identified in each component of G.

The algorithm begins by starting with any vertex v, of G, called the *current vertex*. Any edge e, incident with v is examined. The vertex say u, other than v with which e is incident, becomes the current vertex if it has not yet been numbered. If u has been numbered then e is rejected because, in this case, the acceptance of e would create a cycle. If u has not been numbered then e is accepted as part of the spanning forest and the process is repeated with u replacing v.

Eventually the algorithm will have backtracked through all of the edges in the component of G containing the vertex originally chosen. An unexamined vertex, which will belong to another component (if one exists) is chosen and the process is repeated. When all of the vertices have been numbered, the accepted edges constitute a spanning forest and the algorithm is terminated.

We now illustrate the algorithm by finding a spanning forest of the graph shown in Figure 9.6. The adjacency lists of the graph are illustrated in Table 9.2.

Vertex	Adjacent vertices
1	2, 4, 5
2	1, 3
3	2, 4, 5
4	1, 3
5	1
6	7, 8, 9
7	6, 8
8	6, 7
9	6
10	1

Table 9.2 Adjacency Lists

The initial current vertex in each of the two components are chosen to be vertices 1 and 6. Table 9.3 illustrates the iterations.

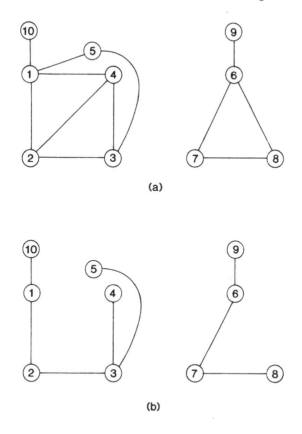

(a)

(b)

Figure 9.6 (a) A graph and (b) its DFS spanning forest

9.5.5 Isomorphism

The *isomorphism problem* of deciding whether two given graphs are isomorphic or not arises frequently in many graph theoretic models, for example in chemistry (see Section 1.1.4 and Section 15.2). The problem of determining whether two given graphs: G_1 and G_2, are isomorphic can be formulated in terms of their adjacency matrices as follows.

Let A_1 and A_2 be the adjacency matrices of G_1 and G_2 respectively. Then $G_1 = (V_1, E_1)$ is isomorphic to $G_2 = (V_2, E_2)$ if a permutation of the rows and columns of A_1 can be found which transforms A_1 into A_2. (This is equivalent to a relabelling of the vertices of G_1 according to the vertex labels of G_2). One can consider testing, at least in theory, all $n!$ possible

Current Vertex, v	$N(v)$	Edge	Acceptance	Spanning Arborescence
1	1			
2	2	$\{1,2\}$	Yes	(1,2)
3	3	$\{2,3\}$	Yes	(2,3)
4	4	$\{3,4\}$	Yes	(3,4)
		$\{4,1\}$	No	
		$\{4,2\}$	No	
3		$\{3,5\}$	Yes	(3,5)
5		$\{5,1\}$	No	
2				
2				
1		$\{1,10\}$	Yes	(1,10)
10	6			
6	7	$\{6,7\}$	Yes	(6,7)
7	8	$\{7,8\}$	Yes	(7,8)
8	9	$\{8,6\}$	No	
7				
6		$\{6,9\}$	Yes	(6,9)
9	10			

Table 9.3 Spanning tree identification.

permutations, where n is the number of vertices of G_1 (and G_2) to discover whether of not G_1 and G_2 are isomorphic. This is clearly impractical for large, arbitrary, pairs of graphs.

Some reduction in the number of permutations that have to be tested can be gained from the following observation. Any one-to-one correspondence between V_1 and V_2, establishing that G_1 and G_2 are isomorphic, will match pairs of vertices with the same degree. Thus only permutations which preserve this property need be considered. Unfortunately, such an approach is still of factorial order and, as such, is of little practical interest.

There is the possibility of a further reduction in the number of permutations to be tested by extending the crude system of vertex classification just discussed, which is based solely on vertex degree. More discerning classifications based on: cliques (maximal complete subgraphs), cycles, the distribution of distance from a given vertex to other vertices, and the dis-

tribution of the degrees of the vertices adjacent to a given vertex have all
been proposed. All such reported algorithms based on this approach are of
exponential order.

A completely different approach to the isomorphism problem is based on
the notion of a *graph invariant* — a property that two isomorphic graphs
either both share or do not share. Possibilities include:

(i) The number of components,

(ii) The number of edges, and vertices of a given degree, and

(iii) The characteristic polynomial

$$\det(\lambda I - A),$$

where A is the adjacency matrix.

Note that any graph invariant is independent of the way in which the graph
is labelled. Unfortunately the characteristic polynomial does not characterize the graph, in the sense that there exist pairs of graphs with the same
characteristic polynomials which are not isomorphic. Attempts to remedy
this situation, by generalizing the characteristic polynomial have not resulted in the reporting of any isomorphic testing algorithm of polynomial
order.

Despite the lack of an algorithm of polynomial order, the isomorphism
problem is not known to be NP-hard. There are efficient algorithms for
the problem when it is restricted to certain classes of graphs:

(i) Graphs with certain special structures (Sussenguth (1965)),

(ii) Rooted trees (Read (1979)),

(iii) Unrooted trees (Aho et al. (1974)),

(iv) Planar graphs (Hopcroft and Tarjan (1972)), and

(v) Graphs without k-strongly regular subgraphs (Corneil (1970)).

9.5.6 Fundamental cycles

In electrical engineering and in other applications of graph theory, it is
sometimes necessary to identify a set of fundamental cycles of a given
graph. This can be done in a straightforward manner by applying the
DFS approach of Section 9.5.1.

Suppose that DFS is employed to examine the edges of a given graph G. In
the course of the search, suppose that an edge e, of G is examined and joins
a labelled vertex to an unlabelled vertex. The addition of this edge to an
arborescence T, of G, (T must have already have been identified) creates
a fundamental cycle C, in G. Note that T is not necessarily a *spanning*
arborescence. Before e is rejected, C is recorded. Upon termination of the

DFS, the cycle C can be retrieved by backtracking. This is achieved by recording, during the search, the *predecessor* $p(v)$, of each vertex. Here $p(v)$ is defined to be the current vertex from which v was labelled. Therefore if an edge $e = v_1 v_2$, joins the current vertex v_1 to a previously labelled vertex v_2, then the fundamental cycle created by e is the vertex sequence:

$$< v_1, p(v_1), p(p(v_1)), \ldots, v_2, v_1 >$$

9.6 Graph optimization algorithms

As we have seen in Section 9.3, any optimization problem can be converted into a decision (yes/no) problem with the use of a bound. Although this does not usually lead to a practical algorithm for solution, it is a useful theoretical device for the analysis of optimization problems, especially those which are intractable. There are a number of graph optimization algorithms which are of importance in many fields where graph theory is applied. These include algorithms for the following:

(i) Shortest paths,

(ii) Minimal spanning trees,

(iii) Maximum weight planar subgraphs, and

(iv) Network flows.

Algorithms for these graph optimization problems, will be explained in Section 12.5. We discuss here the following optimization topics:

(i) The greedy algorithm,

(ii) Colouring, and

(iii) Matchings.

9.6.1 The greedy algorithm

The first optimization algorithm that we describe, the *greedy algorithm*, is the simple-minded strategy of progressively building up a solution, one element at a time, by choosing the best possible element at each iteration.

Consider the following optimization problem. Let Q be a set with weight function $w : Q \to \Re^+$. We define the *weight* of a subset $P \subseteq Q$, to be the sum of the weights of the elements of P. Let ζ be a given class of subsets of Q. Then the problem is to find the member of ζ which has the minimum weight. Naturally, there is a maximization version of the problem which can be defined in an analogous fashion.

There are a number of examples of this problem which are discussed in this book, especially in Part II. We have already seen one such example, the Steiner tree problem, in Section 9.3. The following algorithms can be applied to attempt to solve this problem.

Greedy algorithm statement

Step

1) Choose an element $q \in Q$ where $\{q_1\} \in \zeta$, such that
   $w(q_1) \leq w(q), \forall q \in Q$ with $\{q\} \in \zeta$.
   If such a q_1 does not exist, go to step (4), otherwise continue.

2) Choose an element $q_2 \in Q$ where $\{q_1, q_2\} \in \zeta$, such that
   $w(q_2) \leq w(q), \forall q \neq q_1$ with $\{q_1, q\} \in \zeta$.
   If such a q_2 does not exist go to step (4), otherwise continue.

3) The kth (general) step.
   Choose an element $q_k \in Q$ where $\{q_1, q_2, \ldots, q_k\} \in \zeta$ such that
   $w(q_k) \leq w(q), \forall q \neq q_1, q_2, \ldots, q_{k-1}$, with $\{q_1, q_2, \ldots, q_{k-1}, q\} \in \zeta$.
   If such a q_k does not exist, go to step (4).

4) Terminate.

The set of elements finally chosen by the algorithm is of minimal (but not necessarily minimum) weight in ζ. Thus the algorithm does not guarantee optimality. Conditions for when optimality is guaranteed are given in Section 10.3, along with a numerical example of its implementation. The greedy algorithm is applied to the facilities layout problem in Section 14.2.

9.6.2 Colouring

We introduced in Section 8.2, the idea of colouring the vertices of a graph, so that no two adjacent vertices have the same colour. As will be seen in Part II of this book (for example in Chapter 12 concerning timetabling) the problem of colouring a graph with the minimum number of colours is of practical importance.

We now discuss, in broad terms, some of the available graph colouring techniques. Unfortunately there is no known tight bound for χ, the chromatic number of an arbitrary graph, that is based on simple criteria. Also, there is no reported algorithm of polynomial order for determining χ. In fact the problem of whether or not an arbitrary graph contains a c-colouring with $c \leq k$, for a given k, is NP-hard.

Given this gloomy state of affairs, it is natural to consider colouring heuristic methods which are of polynomial order, but which may require strictly more that χ colours. Such heuristics are the only recourse when a graph that has to be coloured is too large to be coloured in reasonable computing time by a colouring algorithm. We discuss heuristic colouring methods next and then conclude this section with an outline of various colouring algorithms.

One obvious heuristic colouring method begins by arranging the vertices of the graph G, to be coloured, in non-increasing order of vertex degree. The first vertex is coloured with say, colour 1. This vertex and its incident edges are removed from G. The degrees of remaining vertices are recalculated. The list of the remaining vertices is scanned in the order of vertex degree and the first possible vertex that can be coloured with colour 1 is thus coloured. The above steps are repeated until no further vertices can be coloured with colour 1. The process is then repeated with colour 2, then colour 3, and so on, until all vertices are coloured. The colouring produced is termed a *sequential colouring*. Variations of the above methods have been reported by Welsh and Powell (1967), Williams (1968), and Matula et al. (1972) . For many problem instances, sequential colouring heuristics produce colourings which are close approximations to the chromatic number. However they perform arbitrarily badly on certain classes of graphs.

We now turn to colouring algorithms. We begin by sketching a BFS colouring algorithm, due to Christofides (1975), based on the following concepts.

Definitions

A *c-subgraph* of a graph G, is a subgraph of G which is c-chromatic.

A c-subgraph S, is termed *maximal* if S does not contain a proper subgraph which is itself a c-subgraph of S.

Thus a maximal 1-subgraph is an independent subgraph. These notions give us some insight into how to calculate the chromatic number χ, of a given graph $G = (V, E)$. The minimum c for which a maximal c-subgraph of G is spanning is equal to χ. This leads to the following result which is used as a basis for a colouring algorithm.

Theorem 9.7

Any c-chromatic graph G can be coloured with c colours by first colouring the vertices of a maximal independent 1-subgraph of G, say S_1, all with a single colour, then colouring the vertices of another maximal independent 1-subgraph of $G-S_1$, all with another single colour, and continuing in this fashion until all the vertices of G are coloured.

Proof

First we show that a c-colouring of G can always be obtained with only c colours. Because G is c-chromatic, a c-colouring of G exists.

Take a c-colouring of G and assume that one of the vertex sets, say V_1, of G coloured with the same colour, say colour number 1, is not a maximal independent 1-subgraph. We now re-colour with colour number 1 all the vertices of G, say V', which are not in V_1 which form, together with V_1, a maximal independent set in G. The process is now repeated for the graph $G - (V_1 \cup V')$, with a second colour, and so on. This leads to a c-colouring of G of the desired type. ∎

We now use Theorem 9.7 to construct a mechanism for transforming a c-subgraph into a $(c+1)$-subgraph. Let $F_c(G)$ be the collection of all maximal c-subgraphs of G, and $V^j(G)$ be the vertex set of the jth element of F_c. We have seen that a maximal 1-subgraph is an independent subgraph. Thus $V^j(G)$ is the vertex set of a maximal independent subgraph V, of the graph G^j formed by vertices which are not members of the c-subgraph of G induced by $V_c^j(G)$. Thus G^j is induced by $V - V_c^j(G)$. We can identify F_c as follows. Let

p_c be the number of c-subgraphs,

p_1^j, be the number of 1-subgraphs,

$$K^i = V_c^j(G) \cup V_1^k(G^j), \quad j = 1, 2, \ldots, p_c, \ k = 1, 2, \ldots, p_1^j,$$
$$i = 1, 2, \ldots, p_1^j p_c.$$

Then $F_{c+1}(G)$ is equal to the set of maximal K^i's. This gives a basis for finding a χ-colouring of G by generating maximal c-subgraphs, for $c = 1, 2, \ldots$ until a spanning c-subgraph is produced. The colouring of this spanning c-subgraph constitutes the required colouring of G in the minimum number of colours.

DFS colouring algorithms (Wang (1974), Read (1969) , Corneil and Graham (1973), and McDiarmid (1976)) are based on the *chromatic polynomial* $P_k(G)$, of a graph G, which denotes the number of ways of colouring G with k colours. Recall from Section 1.2 that the graph obtained from G by deleting one of its edges is denoted by $G - e$. We define *edge contraction* to be the operation by which two adjacent vertices v_1 and v_2, in a graph G, are merged into one vertex, which is adjacent with all the vertices in G with which v_1 and v_2 were adjacent. If e is the edge joining v_1 and v_2 then the graph resulting from the edge contraction of e is denoted by $G \backslash e$.

This process is illustrated in Figure 9.7(a). It can be shown for any graph $G = (V, E)$ that

$$P_k(G) = P_k(G - e) + P_k(G \backslash e), \forall e \in E.$$

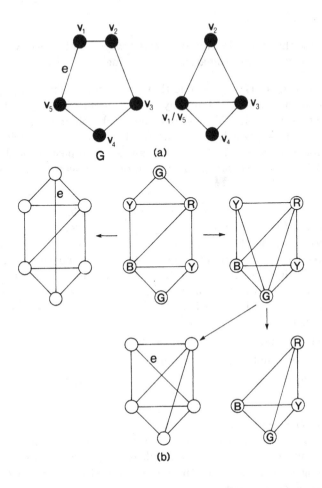

Figure 9.7 (a) Edge contraction

This equation is used to branch in a binary DFS tree. Two new branches are added to the tree at each branching, by (i) adding a new edge and (ii) contracting an edge. When the branching process creates a complete graph, say K_c, then the algorithm can be terminated. In this case a c-colouring can be deduced by backtracking from the node in the tree representing K_c. This is illustrated in Figure 9.7(b). A K_4 has been identified in the leaf of the tree at the bottom right. It is assigned a 4-colouring as shown. A 4-colouring of the original graph is deduced by backtracking from the K_4 by preserving the vertex colours and assigning the same colours to vertices associated by the edges of the contraction process.

9.6.3 Matchings

In Section 8.3 we introduced the notion of a matching in a graph, that is, a subset of its edges in which no two edges are adjacent. We first discuss a maximum matching algorithm due to Edmonds (1965). Recall Theorem 8.7, which states that a matching M, in a graph G, is maximum if and only if G does not have an augmenting path with respect to M. This result is the basis of Edmonds' approach: find a matching M, in the given graph, and an augmenting path P, with respect to M. Create a new matching by replacing the edges in P which are in M by the edges in P which are not in M. This process is repeated, by find an augmenting path for each matching identified.

When all paths are unaugmentable with respect to the current matching, that matching is maximum. The key to Edmonds' algorithm is the efficient identification of each required augmenting path. This is done by using the concept of a *blossom*.

Definitions

A vertex v, is said to be *matched* with respect to a matching M, of a graph if v is incident with an edge of M, and *unmatched* otherwise. A *blossom*, with respect to a matching M in a graph, is an augmenting path with identical initial and final unmatched vertices and an odd number of edges.

We begin our search for an augmenting path P, relative to a matching M, by starting at an unmatched vertex, say v. We must find another unmatched vertex, say v', which is to be the end of P. It often happens that, during the search for a suitable path, an augmenting cycle with an odd number of edges, i.e. a blossom, is identified. As soon as this occurs, all the vertices in the blossom are replaced by one new vertex, called a *pseudo-vertex*. This is achieved by edge contraction of all of the edges of the cycle. The search for a suitable path then continues in the transformed graph. The process of creating pseudo-vertices is repeated if possible. Suppose that an augmenting path in a transformed graph is found. Then an augmenting path P, is identified in G by backtracking from the transformed graph. Each pseudo-vertex is replaced with its blossom. The cardinality of the matching in G is increased by one when P is used and the process is continued. When no further augmenting paths can be found, the current matching is maximum and the algorithm is terminated.

Edmonds' algorithm is $O(n^4)$, where n is the number of vertices in G. This order has been reduced to $O(n^3)$ by Gabow (1976) who has replaced the costly shrinking and expansion of blossoms by using a labelling technique due to Witzgall and Zahn (1965). This has been improved to $O(n^{2.5})$ by Even and Kariv (1975).

Even and Tarjan (1975) have reported another algorithm of $O(n^{2.5})$ for the case when G is bipartite. It departs from the philosoply of the Edmonds-Gabow approach in its choice of an augmenting path. Rather than making an arbitrary choice, Even and Tarjan choose the augmenting path with the smallest number of edges.

9.7 Summary

In this chapter we have completed our survey of graph theoretic concepts by discussing algorithmic graph theory. This will serve as a link between the pure mathematics that we have covered in the earlier chapters, and the applications, which will use the concepts developed so far, especially those of this chapter, in Part II.

Many projects involving graphs, even in pure graph theory itself, involve algorithms. The implementation of these algorithms on nontrivial problem instances requires the use of a digital computer. For this reason we placed heavy emphasis on the computational aspects of graph theory here, especially on the identification of efficient algorithms. Before discussing applications of the concepts developed so far we make a brief excursion into the world of matroids.

9.8 Exercises

9.1 Construct algorithms for transforming inputs (i) – (v) specified in Section 9.2, one into another.

9.2 Suppose that 1800 seconds are available on a computer with which to solve a given problem. Assume, further that two algorithms are available to solve the problem: A_1 with complexity 3^n and A_2 with complexity n^α, where n denotes the size of an instance of the problem. The computer operates by performing 2^{11} elementary operations per second.

  (i) Assume that $\alpha = 8$. What is the maximum problem size which A_2 can solve within the given time?

  (ii) Below what value of n is A_1 more efficient than A_2?

  (iii) What is the maximum problem size which A_1 can solve within the given time?

9.3 Let P_1 be a decision problem for which no polynomial algorithm is known. Suppose that $P_1 \in NP$. Show that an exponentially-bounded deterministic algorithm can be found for P_1.

9.4 Devise a DFS algorithm to find, for a given connected graph G:

  (i) All the spanning trees of G,

(ii) A set of fundamental cut-sets of G,

(iii) Whether G is bipartite, and

(iv) Whether a given subset of the edges of G represents a cut-set of G.

9.5 Devise a polynomially-bounded algorithm to construct a 5-colouring of any planar graph.

9.6 Devise an algorithm for generating the adjacency matrix of a *random graph*. That is, a graph with a given number n, of vertices, whose edges are drawn at random according to a fixed given probability, from the set of all edges in K_n.

9.7 Devise an algorithm for converting the final matrix produced by the connectivity algorithm (Section 9.5.2) back to the adjacency matrix.

9.8 As defined in Section 2.1, a *bridge* in a connected graph G, is an edge whose removal from G, renders G no longer connected. Devise a DFS algorithm to find all the bridges of a given connected graph.

9.9 Devise a DFS algorithm to find all the blocks of a given connected graph.

9.10 Devise a DFS algorithm to find a minimum colouring of a given graph.

9.11 Show that the problem of identifying the maximum-weight planar subgraph of a given weighted graph is NP-hard. Hint: Use the fact that the following problem is NP-hard. Given a graph (V, E), and a positive integer $K \leq |E|$, determine whether or not there exists a subset $E' \subseteq E$, with $|E| \geq K$ such that the graph (V, E') is planar.

10 Matroids

I am a little world made cunningly of elements.

Holy Sonnets, Annunciation — John Donne

Consider a finite set of vectors over an arbitrary field. Of course, each subset of this set is either linearly independent or linearly dependent. Also, any subset of a linearly independent subset of the original set is itself linearly independent. Further, if I and J are two linearly independent subsets of the original subset with $|I| = |J| + 1$, then there exists an element of I which, together with J, forms a linearly independent set of $|I|$ vectors.

These properties are also possessed by algebraic systems other than the one just mentioned. For instance: (i) the collection of the subsets of edges of a graph which do not contain any cycle, and (ii) the collection of subsets of edges of a graph which do not contain any cut-set. Whitney (1935) was the first to popularize the notion of a *matroid*, which is a set with an independence structure defined on it, as just illustrated.

We shall introduce some fundamental concepts relating to matroids. We shall see that a deeper understanding of the notion of duality in graph theory can be gained by studying matroid structures. Matroids have practical application in operations research, and in other fields of study where one wishes to apply optimization algorithms to graph theoretic models, especially the so-called *greedy algorithm*, as explained in Section 9.6.1. We begin by motivating the introduction to matroids via various graph theoretic concepts.

10.1 Introduction

Recall that in Section 3.6 we defined a spanning tree in a connected graph G, to be an acyclic subgraph containing all of the vertices of G. Obviously no proper subgraph (containing a proper subset of the set of vertices) of a spanning tree of G is itself a spanning tree of G.

Suppose that T_1 and T_2 are two distinct spanning trees of a graph G. Let e_1 be any edge of T_1. Then it is always possible to find an edge, say e_2, in T_2 but not in T_1, such that the graph obtained by replacing e_1 by e_2 in T is also a spanning tree of G. An analogous result holds for any two distinct bases of vector space.

Definition

A *matroid* $M = (E, \zeta)$, is an ordered pair where E is a finite nonempty set of *elements* and ζ is a collection of subsets of E, whose members are called *independent sets* of M where:

M(i) $\phi \in \zeta$,

M(ii)   If $I \in \zeta$ and $J \subset I$, then $J \in \zeta$, and

M(iii)   If I and $J \in \zeta$ with $|I| = |J| + 1$, then there exists an element $e \in I \backslash J$, such that $(J \cup \{e\}) \in \zeta$.

Definition

A maximal independent set of M, that is an independent set not contained in a larger independent set of M, is called a *base* of M.

The axioms can be used to show that all the bases of any matroid M, possess the same number of elements, which is called the *rank* of M and is denoted by $r(M)$.

There are three matroids $M_1 = (E_1, \zeta_1)$, $M_2 = (E_2, \zeta_2)$, and $M_3 = (E_3, \zeta_3)$, which arise naturally from any graph $G = (V, E)$:

(a) M_1 the *graphic* or *cycle matroid* of G:

Let $E_1 = E$ and ζ_1 be the collection of all subsets of E which do not contain any cycles in G.

(b) M_2, the *co-graphic* or *cut-set matroid* of G:

Let $E_2 = E$ and ζ_2 be the collection of all subsets of E which do not contain any cut-set in G.

(c) M_3, the *matching matroid* of G:

Let $E_3 = V$ and ζ_3 be the collection of all subsets V' of V, such that the elements of V' are matched in some matching in G.

It can be seen that any matroid can be completely specified by stating its independent sets.

Definitions

If $M = (E, \zeta)$, is a matroid then a set $E' \subseteq E$ is said to be *dependent* if E' does not belong to ζ.

A minimal dependent set is termed a *cycle*.

It appears that this last definition is well chosen as the cycles in matroid M (just introduced) arising from a graph G, are exactly the cycles of G. Further, the realization that $E' \subseteq E$ is independent, if and only if it corresponds to an acyclic subgraph, leads to the following alternative definition of a matroid.

A *matroid* $M = (E, \Omega)$ is an ordered pair where E is a finite nonempty set of *elements* and Ω is a nonempty collection of subsets of E called *cycles*, where:

AM(i) No proper subset of a cycle is a cycle, and

AM(ii) If $C_1, C_2 \in \Omega$ and $c \in C_1 \cap C_2$ then there exists a cycle in $C_1 \cup C_2$ which does not include c.

If $M = (E, \zeta)$ is a matroid and $E' \subseteq E$, then the size of the largest independent set contained in E' is called the *rank* of E' and is denoted by $r(E')$.

Thus if $M = (E, \zeta)$, is a matroid, then $r(M) = r(E)$.

As we have seen, we can associate a matroid M, with the edge set of any graph G, by defining the cycles of M to be the cycles of G. We now ask whether the reverse process is well-defined. Given any matroid M, does there exist a graph whose cycles correspond to the cycles of M? Matroids for which the answer is *yes* are termed *graphic matroids*. So, are all matroids graphic? The answer is *no*. As an example of a matroid $M = (E, \zeta)$ that is not graphic, let $|E| = 4$, and let the bases of M be the subsets of E containing exactly two elements. Then the independent sets of ζ are those subsets of E containing not more than two elements. Thus if $E = \{e_1, e_2, e_3, e_4\}$ we have:

Bases   : $\{e_1, e_2\}, \{e_1, e_3\}, \{e_1, e_4\}, \{e_2, e_3\}, \{e_2, e_4\}, \{e_3, e_4\}$.

Independent Sets : $\emptyset, \{e_1\}, \{e_2\}, \{e_3\}, \{e_4\}, \{e_1, e_2\}, \{e_1, e_4\},$
                        $\{e_3, e_4\}, \{e_1, e_3\}, \{e_2, e_3\}, \{e_2, e_4\}.$

Definition

A matroid $M = (E_1, \zeta)$, whose bases are those subsets of E with exactly k elements is termed k-*uniform*.

Thus the nongraphic matroid just discussed is 2-uniform.

We have introduced graphic matroids, but it is now time for rigorous definitions.

Definitions

Two matroids $M_1 = (E_1, \zeta_1)$ and $M_2 = (E_2, \zeta_2)$ are said to be *isomorphic* if there exists a one-to-one correspondence between the elements of E_1 and of E_2 which preserves *independence*.

A matroid is said to be *graphic* if·it is isomorphic to the graphic matroid of some graph.

A matroid is said to be *co-graphic* if it is isomorphic to the co-graphic matroid of some graph.

A matroid is said to be *planar* if it is both graphic and co-graphic.

A matroid is said to be *bipartite* if each of its cycles contains an even number of elements.

A matroid M, is said to be *Eulerian* if its set of elements is the union of a set of disjoint cycles of M.

10.2 Duality

We shall now study one area of the theory of matroids which has strong links with graph theory, namely *duality*. We shall show that graph theoretic notions of duality are more intuitively appealing when the corresponding matroid notions are examined. Thus a study of matroid duality reveals a generalization of graph theoretic duality and makes the matter seem more natural and easier to understand.

We saw in Theorem 5.10 that the set of edges in a planar graph G, constitutes a cycle if and only if the corresponding set of edges in its dual constitutes a cut-set. We can form both a graphic and a co-graphic matroid from G by taking as the minimal dependent sets: (a) the cycles of G and (b) the cut-sets of G respectively. Given the relationship between a graph and its dual, it is natural to define matroid duality in a similar way.

Definition

If $M = (E, \zeta)$, is a matroid with a set of bases β, then the matroid with a set of elements E, and a set of bases $\beta^* = \{S - B \mid B \in \beta\}$ is termed the *dual* of M and is denoted by M^*. That M^* is a matroid is stated in Theorem 10.1.

It can easily be shown that $(M^*)^* = M$ and thus M and M^* are referred to as a *dual matroid pair*.

Definitions

The set of elements of a matroid M form a

Co-cycle of M if they form a cycle of M^*, and a

Co-base of M if they form a base of M^*.

There are other features of M with a "co" prefix corresponding to the same feature of M^*.

We denote the rank function of M^* by r^* and state without proof some results concerning matroid duality.

Theorem 10.1

If M is a matroid then M^* is a matroid.

Theorem 10.2

If M and M^* are a dual matroid pair with element set E, then for all $E' \subseteq E$, $r^*(E') = |E'| + r(E - E') - r(E)$.

Corollary 10.1

If $M = (E, \zeta)$ is a matroid, and $E' \subseteq E$ is independent in M, then $E - E'$ contains a co-base of M.

Corollary 10.2

If M^* is the dual of a matroid $M = (E, \zeta)$ and $E^* \subseteq E$ is independent in M, then $E \backslash E^*$ contains a base of M.

Theorem 10.3

Every co-circuit of a matroid M has an element in common with every base of M.

Corollary 10.3

Every cycle of a matroid M, has an element in common with every co-base of M.

10.3 The Greedy Algorithm

Let $G = (V, E)$ be a weighted connected graph with weight function $W : E \to \Re^+$.

We define the weight of a subgraph G' of G to be the sum of the weights of the edges of G'. As we shall see later it is often of interest to attempt to identify a subgraph of G with a given property, that has either maximum or minimum weight. Examples of the given property are, *is a spanning tree, is planar, is an Euler trail,* or *is a Hamiltonian cycle.*

We can begin to analyse such combinatorial optimization problems on the graph G, by recalling the *greedy algorithm,* defined in Section 9.6.1. We discovered that there is no guarantee that the algorithm will identify the optimal subgraph. As an example of a suboptimal performance of the greedy algorithm, consider the problem of finding the shortest path from vertex 1 to vertex 4 in the graph in Figure 10.1, with the edge weights given. The greedy algorithm, when applied to this problem will identify the path, $\langle 1, 2, 3, 4 \rangle$. However the shortest path is $\langle 1, 3, 4 \rangle$. But when w_{23} is reduced to 18, the greedy algorithm will again identify the path $\langle 1, 2, 3, 4 \rangle$, which is now the shortest.

We shall now examine the greedy algorithm in terms of matroid theory. Let $M = (E, \zeta)$, be a matroid with a weight function $W : E \to \Re^+$. Let I and J be members of ζ where:

$$I = \{p_1, p_2, \ldots, p_n\}, \quad \text{and}$$
$$J = \{q_1, q_2, \ldots, q_m\}, \quad \text{where}$$
$$w(p_1) \le w(p_2) \le \ldots \le w(p_n), \quad \text{and} \tag{10.1}$$
$$w(q_1) \le w(q_2) \le \ldots \le w(q_m). \tag{10.2}$$

Definitions

I is said to be *lexicographically smaller* then J if there exists an integer k such that $w(p_i) = w(q_i)$ for $i = 1, 2, \ldots, k - 1$ and either:

(i) $w(p_k) < w(q_k)$, or

(ii) $w(p_i) = w(q_i)$, for $i = 1, 2, \ldots, n$ and $n < m$.

If no independent set is lexicographically smaller than an independent set I, then I is said to be *lexicographically minimum.*

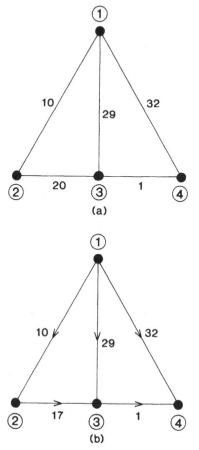

Figure 10.1 (a) A weighted graph and
(b) a weighted digraph

Analogous definitions can be stated to define a *lexicographically maximum* set.

From the above it follows that a lexicographically minimum set is a base of M.

A set B in ζ is *Gale optimal* if, for every I in ζ, there exists a one-to-one correspondence between I and B such that, for all $p \in I$,

$$w(p) \geq w(q), \qquad (10.3)$$

where q is the element of B which is mapped to p by the one-to-one correspondence.

We can make some observations:

GO(i)   *Gale optimal* can be defined in terms of *maximality* by reversing the inequality sign in (10.3).

GO(ii)   Only bases can be Gale optimal.

GO(iii) If a base is Gale optimal it must have minimum weight (maximum weight if the definitions corresponding to maximization have been assumed).

These observations lead us to the following useful theorem:

Theorem 10.4

Let $M = (E', \zeta)$ be a matroid having nonnegative weights associated with its elements arranged in the order of (10.1), with base B. Then

GO(i) B is lexicographically minimum if and only if,

GO(ii) B is Gale optimal, if and only if,

GO(iii) B is a minimum weight member of ζ.

Theorem 10.4 implies that the greedy algorithm will identify a lexicographically minimum base which has minimum weight. This leads us to:

Theorem 10.5

Let $M = (E, \zeta)$, be a matroid having nonnegative weights associated with its elements. The greedy algorithm, when applied to ζ, will identify a minimum weight member of ζ.

Naturally there exist theorems analogous to Theorems 10.4 and 10.5 for the case of maximization. From the above results it is obvious that we shall identify a minimum (maximum) weight base if we choose the elements of S in the order of (10.1) (order which is the reverse of (10.1)). The only proviso is that an element must be passed over in this selection process if its addition would create dependence in the subset of elements chosen so far. A greedy method for the minimum spanning tree problem is stated and illustrated in Section 12.5.1. In view of the previous development, the reader may be wondering why the greedy algorithm failed to identify the shortest path between vertices 1 and 4 in the graph in Figure 10.1. The following theorem provides conditions for when the greedy algorithm will succeed.

Theorem 10.6

Let ζ be the collection of subsets of a set E where, if $I \in \zeta$ and $J \subseteq I$ then $J \in \zeta$. Then for all nonnegative weight functions defined on the elements of E the greedy algorithm, when applied to ζ, will identify the minimum weight member of ζ only if (E, ζ) is a matroid.

Proof

Exercise.

Theorem 10.6 implies that a sufficient condition to guarantee that the greedy algorithm will identify the minimum weight member, is that ζ is the collection of independent sets of a matroid on S. Naturally a maximization version of Theorem 10.6 can be stated analogously.

10.4 Summary

In this chapter we have given a skeleton outline of the fundamental concepts of the theory of matroids. Our development of this intricate area of mathematics has been, mainly due to space limitations, terse and related only to the relationship with graph theory and the greedy algorithm used in Chapter 12. For a fuller and more general explanation of matroids the reader is referred to Welsh (1976) and von Randow (1975) , and to Lawler (1976) for applications of matroids.

10.5 Exercises

10.1 Prove that the two definitions of a matroid stated in Section 10.1 are equivalent.

10.2 Show that M_1, M_2, and M_3, defined in Section 10.1 are matroids.

10.3 Show that the cycle matroid and the cut-set matroid of any graph are a dual matroid pair.

10.4 Let I_2 be the field of integers modulo two. If a matroid M, is isomorphic to a vector matroid defined in some vector space of I_2, then M is said to be *binary*. Show that M is a binary Eulerian matroid if and only if M^* is a binary bipartite matroid.

10.5 Prove all observations and theorems stated in this chapter.

10.6 Prove that when the greedy algorithm has identified m elements that these have minimum total weight compared to all independent sets with m or fewer elements.

Part II: Applications

The bearings of this observation lay in the application on it.

Dombey and Son, Charles Dickens.

We come now to the main part of this book where it is shown how the graph theory introduced in Part I can be applied in a wide variety of fields. It has been seen in Part I that graph theory is an elegant branch of pure mathematics, of interest in its own right. The purpose of Part II is to show that the theory has power in the analysis of phenomena and problems from a wide variety of natural and societal problems.

11 Miscellaneous Applications

Variety's the very spice of life.

The Task, William Cowper

In Part I of this book we have introduced the theory of graphs with an occasional application of that theory. Examples include the counting of evolutionary trees in biology (Section 3.5), the intractability of optimizing phylogenies (Section 9.3), problems in tournaments (Section 7.6), and in chemistry and physics (Sections 1.4 and 1.5). In the following chapters we shall discuss in some depth the application of graph theory in a few selected areas: operations research (Chapter 12), electrical engineering (Chapter 13), industrial engineering (Chapter 14), science (Chapter 15), and civil engineering (Chapter 16).

Some of the reported applications of graph theory in other fields are given below. This list is, for reasons of space, limited in its depth and scope. There are numerous applications in other areas and many other reports equivalent to those mentioned. The reader is urged to use these references as signposts to the rest of the wealth of literature on the utility of graph theory.

1. **The Social Sciences**

   (Roberts (1978), Busacker and Saaty (1965) , Marshall (1971), and

Harary et al. (1965).)

Indifference, measurement, seriation, food, webs, niche overlap, boxicity, scheduling of service vehicles and meetings, communications, location, voting, balance and social inequity, group stability, urban studies, consistency of choice, ranking, and organizational structure. (See also Chapter 12).

2. Economics and Logistics

(Avondo-Bodino (1979), Busacker and Saaty (1965).)

Input-output models.

3. Geography

(Cliff, Haggert and Ord (1979).)

Stream analysis, transportation networks, and market structures. (See also Chapter 12.)

4. Architecture

(Earl and March (1979).)

Floor plan arrangements, enumeration, fundamental plans, ornamentation, region shape, and dimension. (See also Chapter 14.)

5. Puzzles and Games

(Busacker and Saaty (1965), Tucker (1980).)

System transitions, crossing, board games, and coloured cubes.

6. Computer Science

(Deo (1974).)

Information retrieval, program analysis, and optimization. (See also Chapters 2, 9, and 12.)

Graph theory and other branches of mathematics have been highly successful in certain academic fields such as the natural sciences and engineering. This is because of the presence of a well-developed structure in those fields to which mathematical techniques can be applied. In less structured areas of human endeavour, such techniques cannot be applied in the same manner because such structures have often not been developed. However graph theoretic models can sometimes provide a useful structure upon which analytical techniques can be used.

We now choose a small representative sample of problems from outside the natural sciences and engineering, in order to demonstrate the possibilities of graph theory in other academic fields.

11.1 The social sciences (Based on Foulds (1989).)

In this section we present a representative sample of the application of digraphs in the social sciences. Typically a digraph, or graph, is used to model the relationships between a given set of objects. Each object is represented by a vertex, and the relationship between various pairs of objects is represented by one of:

(i) an arc, if the relationship is ordered (e.g. is **taller than**), or

(ii) an edge, if the relationship is unordered (e.g. is **married to**).

Further the arcs or edges can be *signed* by assigning each of them one of two symbols usually "+" and "-'. These signs can be used to indicate a different type of relationship, such as: allied with/allied against, attracted to/repelled by, and so on.

11.1.1 Organizational structure

Consider an organization whose members and rankings are represented by the vertices and arcs respectively of a digraph. This abstraction of the *corporate structure chart* will contain an arc (u, v), if and only if the individual represented by vertex v is inferior (within the organization structure) to the individual represented by vertex u.

The social science concept of *social status* can be quantified with such a digraph. For most organizations involving human beings, the associated digraph is acyclic, as individuals cannot be their own superiors. (Cycles may occur in digraphs modelling the organizational structures of certain flocks of birds which have circular pecking orders.) The level of subordinacy can be established for any individual, v say, in the organization by calculating the length (in terms of the digraph) of the number of arcs of the shortest path from the vertex representing the superior to the vertex representing v.

A collection of vertices with the same level of subordinacy represents a peer group of individuals. It is possible to quantify the *status* of an individual by defining a measure based on the number of subordinates the individual has at each possible level. Efficient shortest path algorithms for calculating the required paths are well-known . Some of these are discussed in Section 12.5.2.

11.1.2 Social hierarchy

Digraphs are often used to model kinship in anthropological groups. As an example, consider a group of individuals who are represented by the vertices of a digraph D which contains an arc (u, v) if and only if the individual represented by vertex u is a parent of the individual represented by vertex v.

Suppose that a second digraph D_2, is similarly constructed, which models the parent-child relationships of individuals which are children in D_1.

In D_2, all possible relationships involving the children of these individuals are modelled. The product of the adjacency matrices of D_1 and D_2 is the adjacency matrix of a further digraph which models the grandparent-grandchild relationships between these individuals.

11.1.3 Consistency of choice

In many social science experiments, one is asked to rank a number of given objects by pairwise comparison, especially when numerical measurement is difficult. For instance, a dog may be presented with each different pair of dog foods from a set of n foods, one pair per day. The food out of the pair that the dog selects first is noted. The object is to rank the n foods in order of preference. The outcome of each day's selection can be modelled as a tournament digraph D, in which each food is represented by a vertex and an arc (u, v), is present if and only if food u is preferred to food v. If D is acyclic, a clear order of preference can be established. If D is cyclic, digraph techniques involving ranking with the minimum number of violations can be employed to establish the best order according to certain optimality criteria. This application of tournaments is discussed in Section 7.6.

11.1.4 Conclusions

For many decades, the theory of digraphs has provided useful tools for areas of the natural sciences where sometimes other branches of mathematics have been of less use. The success of digraphs can, in part, be attributed to the special physical structures present in the problems analysed. For a lesser time it has been apparent that the same digraph modelling approach can be applied with fruitful results in the less structured, social sciences.

One of the major contributions of digraph theory has been the provision of precise, mathematically-based language and concepts by which the social scientist can formulate, hypothesize, conjecture, communicate, and analyze societal problems. What has not happened, to any great extent, is the gaining of major breakthroughs in social science by the application of the mathematical theory of digraphs.

However, as more use of the language and techniques of digraphs becomes prevalent, social scientists will be able to analyze their problems and develop more accurate theories. Conversely, the just-mentioned endeavours have lead to many important results in digraph theory and to new unsolved theoretical digraph problems.

11.2 Economics (Based on Avondo-Bodino (1979).)

In this section we consider the *Leontief open input-output economic model.* It can be used to predict the amounts of commodities which must be produced by the industrial sectors of an economic system in order to satisfy the total demand of its economy. We make a number of assumptions about the system:

(i) The quantities demanded are known.

(ii) In a state of equilibrium, the quantity of each commodity produced by each industrial sector is totally consumed by the other sectors and the final demand.

Formally, let

Q_i be the quantity of commodities produced by the ith industrial sector, termed the ith *net output,*

q_{ij} be the quantity of the commodities produced by the ith industrial sector that is consumed by the jth industrial sector (assumed to be positive),

d_i be the quantity of the commodities produced by the ith industrial sector that is consumed by the final demand, and

n be the number of industrial sectors.

Then

$$Q_i = \sum_{j=1}^{n} q_{ij} + d_i, \qquad i = 1, 2, \ldots, n \qquad (11.1)$$

(iii) The total value of the commodities consumed by the jth industrial sector equals the total value of the commodities made by the jth industrial sector less the added value. Formally, let

p_i be the price of the ith commodity (assumed to be positive), and

v_j be the value added by the jth industrial sector.

Then

$$\sum_{i=1}^{n} q_{ij} p_i + v_i = Q_j p_j, \qquad j = 1, 2, \ldots, n. \qquad (11.2)$$

(iv) There are no economies or diseconomies of scale in the quantity, say a_{ij}, of the commodities produced by the ith industrial sector required to produce units of the commodities consumed by the jth industrial sector. Formally,

$$a_{ij}Q_j = q_{ij}, \qquad i = 1, 2, \ldots, n.$$

We set

$$a_{ii} = 0, \qquad i = 1, 2, \ldots, n.$$

Substituting these values of q_{ij} in (11.1) and (11.2), we have

$$Q_i = \sum_{j=1}^{n} a_{ij}Q_j + d_i, \qquad i = 1, 2, \ldots, n, \tag{11.3}$$

and

$$\sum_{i=1}^{n} a_i Q_j p_i + v_j = Q_j p_j, \qquad j = 1, 2, \ldots, n. \tag{11.4}$$

We now express (11.3) and (11.4) in more compact form by introducing the following matrix notation. Let

$$Q = (Q_i)_{1 \times n}, \qquad D = (d_i)_{1 \times n}, \qquad A = (a_{ij})_{n \times n}, \qquad P = (p_i)_{1 \times n},$$
and $V = (v_j)_{1 \times n}.$

Then (11.3) and (11.4) can be expressed, respectively, as

$$(I - A)Q = D, \tag{11.5}$$

and

$$(I - A)^T Q = W,$$

where

$$W = (w_i)_{i \times n},$$

and

$$w_i = v_i/p_i, \qquad i = 1, 2, \ldots, n.$$

It is natural to ask when solutions for Q exist for (11.5), and when, if the answer is in the affirmative, there is a unique, meaningful, nonnegative solution.

Because the dimension n, of the above system defined by (11.5), is quite large in practice, it is usually inefficient to employ standard algebraic techniques to solve the system. It is usually much more efficient to employ a digraph model as follows.

Some conditions on the existence and nature of solutions to (11.5) are known from matrix theory. Firstly, (11.5) will have a unique solution if and only if

$$\det(I - A) \neq 0.$$

Secondly, if

$$\begin{aligned}
a_{ij} &> 0, & i &\neq j, \\
a_{ii} &< 1, & i &= 1, 2, \ldots, n, \\
d_i &> 0, & i &= 1, 2, \ldots, n, \quad \text{and} \\
w_i &> 0, & i &= 1, 2, \ldots, n,
\end{aligned}$$

then a positive solution to (11.5) exists if and only if the principal minors of $(I - A)$ are positive.

Thirdly, $(I - A)^{-1}$ has positive entries if and only if the eigenvalues of A have magnitude between -1 and $+1$.

This assumes that A is irreducible and can be expressed in block diagonal form. Let

$c_{ij} =$ the consumption, in terms of funds flowing from the ith industrial sector to the jth industrial sector due to purchases of commodities, for $i = 1, 2, \ldots, n$ and $j = 1, 2, \ldots, n,$

$c_{0j} =$ the flow (in the above terms) from final consumer demand to the jth industrial sector, and

$c_{i0} = 0, \quad i = 0, 1, 2, \ldots, n.$

Then the matrix $C = (c_{ij})_{(n+1) \times (n+1)}$ is termed a *consumption matrix*. It is apparent that

$$\begin{aligned}
c_{0j} &= d_j p_{ij} & j &= 1, 2, \ldots, n, \\
c_{ij} &= q_{ji} d_i, & i &= 1, 2, \ldots, n, \ j = 1, 2, \ldots, n, \ i \neq j, \quad \text{and} \\
c_{ii} &= 0, & i &= 1, 2, \ldots, n.
\end{aligned}$$

Further, define

$$\begin{aligned}
b_{ii} &= Q_i p_i, & i &= 1, 2, \ldots, n, \\
b_{ij} &= 0, & i &= 1, 2, \ldots, n, \ j = 1, 2, \ldots, n, \ i \neq j, \quad \text{and} \\
b_{00} &= 0.
\end{aligned}$$

Finally, define

$$e_{ij} = b_{ij} - c_{ij}, \quad i = 1, 2, \ldots, n, j = 1, 2, \ldots, n.$$

It is clear that

$$e_{0j} = -d_j p_j, \qquad j = 1, 2, \ldots, n,$$
$$e_{ii} = Q_i p_i, \qquad i = 1, 2, \ldots, n,$$
$$e_{ij} = -q_{ji} d_i, \qquad i = 1, 2, \ldots, n, \ j = 1, 2, \ldots, n, \ i \neq j, \qquad \text{and}$$
$$e_{i0} = 0, \qquad i = 0, 1, 2, \ldots, n.$$

Let E be the matrix $(e_{ij})_{(n+1) \times (n+1)}$. By using (11.3), we can show that the (1,1) cofactor of E is

$$\det E = (\prod_{i=1}^{n} p_i)(\prod_{j=1}^{n} Q_i) \det(I - A)^T.$$

Hence if $\det E_{11}$ is nonzero then $\det(I - A)$ is nonzero. We can use the theory of digraphs to determine whether $\det E_{11}$ is nonzero. We do this by constructing a digraph $D = (U, A)$, where $U = \{u_0, u_1, \ldots, u_n\}$, with u_0 representing the final (consumer) demand and u_i representing the ith industrial sector, $i = 1, 2, \ldots, n$.

Each arc $u_i u_j \in A$, represents the net flow of funds from the ith industrial sector, or from final demand, to the jth industrial sector due to purchases of commodities. Then, due to a theorem of Bott and Mayberry (1954) we can deduce that if D has a *spanning arborescence* (defined in Section 7.3) rooted at u_0, then $\det E_{11}$ is positive. In this case $\det(I - A)$ is positive, because p_i and Q_j are assumed to be positive for all i and j.

It is usually more convenient computationally, to identify a spanning arborescence than to solve (11.5) for Q.

As an example, let $n = 3$, and suppose that the flow of funds is given by:-

$$C = \begin{array}{c} \\ 0 \\ 1 \\ 2 \\ 3 \end{array} \begin{array}{c} \begin{array}{cccc} 0 & 1 & 2 & 3 \end{array} \\ \left[\begin{array}{cccc} 0 & 2 & 8 & 6 \\ 0 & 0 & 0 & 3 \\ 0 & 4 & 0 & 0 \\ 0 & 0 & 7 & 0 \end{array} \right] \end{array}.$$

An equivalent Leontif digraph is given in Figure 11.1. All the spanning arborescences of this digraph which are rooted at u_0 are shown in Figure 11.2.

We define the weight of each arborescence to be the product of its arc weights. The above mentioned theorem states that $\det E_{11}$ equals the sum of the weights of all the spanning arborescences of D. Thus as long as D has at least one spanning arborescence rooted at u_0 then $\det E_{11} > 0$.

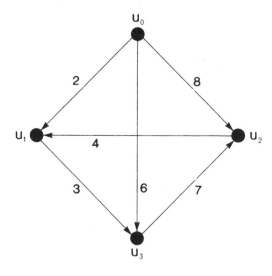

Figure 11.1 A Leontif digraph

11.3 Geography (Based on Cliff et al. (1979).)

The Four Colour Theorem, mentioned in Sections 1.1.3 and 8.2, has relevance to social geography concerning, for example, periodic market day analysis (Tinkler (1973)). Transportation networks (see Section 12.5.3) are also of interest in geography.

In this section we demonstrate an application of graph theory in physical geography, namely in analyzing the topological structure of stream networks. This work was begun by Strahler (1964), Shreve (1966) , and Smart (1967, 1969).

11.3.1 A Hierarchical graph model of stream networks

Consider a network of streams which does not contain anastomosing sections (i.e. loops). The topology of such a network can be modelled by an aborescence (see Section 7.3), rooted at the ultimate stream sink. Such models have been used by geographers to analyse the way in which stream systems evolve and also their propensity to flood.

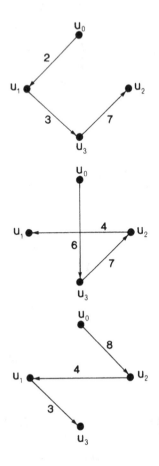

Figure 11.2(a) The first three spanning arborescences
of the Leontif digraph

We now develop the notion of *stream order* (Strahler number) in which each
stream segment is numbered as follows. Initially, all stream channels are
not numbered. Each initial stream source (corresponding to a pendant arc
in the aborescence) is numbered one. Each unlabelled stream channel (arc)
adjacent to a numbered stream segment (arc) is assigned a number equal to
one plus the maximum of the numbers of the numbered channels (arcs) with
which it is adjacent. Thus the channels (arcs) are progressively numbered
until finally the channel (arc) incident with the root is numbered. This
process is illustrated in Figure 11.3 (a), in which a stream system is shown

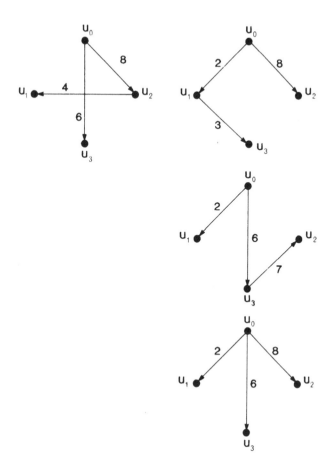

Figure 11.2 (b) The four remaining spanning arborescences
of the Leontif digraph

and (b) in which the arcs of the associated aborescence have been numbered
(ignoring the values in paretheses) by the above-mentioned process.

The *order* of each stream channel is defined to be its assigned number.
Thus the system in Figure 11.3 is of the seventh order.

Let

  n_i be the number of streams of order i, and

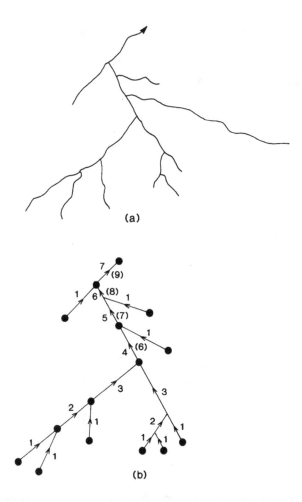

Figure 11.3 (a) A stream network and (b) its associated
hierarchically numbered arborescence

r_i be the *bifurcation ratio* defined as

$$r_i = \frac{n_i}{n_{i+1}}, \qquad i = 1, 2, \ldots \quad .$$

For the network in Figure 11.3, we can calculate these ratios as:

$$r_1 = \frac{8}{2} = 4, \qquad r_2 = \frac{2}{2} = 1, \qquad r_3 = \frac{2}{1} = 2,$$

$$r_4 = \frac{1}{1} = 1, \qquad r_5 = \frac{1}{1} = 1, \qquad \text{and} \qquad r_6 = \frac{1}{1} = 1.$$

The average of all the bifurcation ratios and the relationship between n_i and i, $i = 1, 2, \ldots$, can be calculated. These parameters are used to predict the elapse of time between the occurrence of storm rain in a stream network and its consequent flooding. It appears that the lower the bifurcation ratio, the more serious the flood (Kirkby (1976)).

11.3.2 A random graph model of steam networks

We turn now to a second numbering system for the channels of stream network. In this process each channel is assigned a *magnitude* (rather than an order) equal to the number of sources upstream from it. Once again, the numbers for any network can be calculated sequentially, first by assigning a unit magnitude to each pendant channel (arc) and then progressively calculating magnitudes for downstream channels. This has been shown in Figure 11.3 (b) for the system in Figure 11.3 (a), where magnitudes that differ from orders have been shown in parentheses. Thus this system is of the ninth magnitude. As before, the channel magnitudes of a stream network can be used to calculate parameters of the network from which deductions are made about its geographical properties.

The aborescences of the set of stream networks with a given number of sources belong to a number of isomorphism equivalence classes. Assume that every possible aborescence of a given set is equally likely to arise in nature. Then it is interesting to ask: (i) what is the most likely order to occur? and, (ii) does this order actually occur in nature with the frequency predicted by this random model? For instance, Werrity (1972) has demonstrated that, for a stream network with 50 sources, a fourth-order system is most likely with an average bifurcation ratio of 3.83. It appears that in nature most stream systems have an average bifurcation ratio of approximately 4.0, and that significant departures from this norm usually occur as a result of human activity.

11.4 Architecture

The concept of graph planarity, introduced in Chapter 5, has relevance in facilities layout, to be discussed in Chapter 14. This notion is also of interest in architecture. In this section we provide a glimpse of some of the ideas to be examined in depth in that later chapter.

One of the central questions in architectural design concerns the identification of suitable floor plans representing the possible designs of some system. With the advent of the digital computer, the designer is now able to enumerate implicitly many possible plans for design projects. We shall illustrate how graph theoretic concepts can aid the enumerative process for one particular design problem.

11.4.1 Floor plan design enumeration

In this section we enumerate the number of feasible representations of the solutions to a given floor plan design problem. The topic of floor plan design is explored in greater depth in Section 14.2. Another illustration of graph enumeration is given in Section 3.5. An alternative treatment of this topic, and other applications of graph theory in architecture, have been given by Earl and March (1979). We shall assume that a single rectangular floor of a future building is to be designed. The building will be used to house the facilities or activities of some system which is to occupy the floor. Each facility can be represented by a rectangle of given area to be located within the floor with its walls parallel to the outer walls of the floor. The sum of the areas of the facilities equals the area of the floor.

The *floor plan design problem* involves finding a tesselation of the floor rectangle into areas representing the given facilities, according to the above assumptions, which is optimal according to some given criteria.

If one of the optimality criteria is related to which pairs of facilities are adjacent, the notion of graph planarity is of use in the above problem. This is because we can represent which pairs of facilities are adjacent in a given floor plan by means of an appropriate planar graph, with its vertices and edges representing facilities and adjacencies respectively.

We shall enumerate all feasible (planar) adjacency graphs for the layouts of n given facilities. In graph theoretic terms, we address the question of enumerating the maximal planar subgraphs of a vertex-labelled K_n. The following is a key result in this endeavour. It was proved as Equation (1.5) on page 22 of Tutte (1962).

$$\psi_{n,0} = \frac{2(3n+3)(3n+4)\ldots(4n+1)}{(n+1)!}, \quad n \geq 2, \qquad (11.6)$$

where $\psi_{n,m}$ is the number of nonisomorphic triangulations of a polygon with k given external edges and r internal edges. Here $k = (m+3)$ and $r = (3n+m)$. For a triangulation of the sphere with a rooted region (which can be thought of as external for the polygon), we have $k = 3$, and hence $m = 0$ and $r = 3n$. The boundary of this rooted region has vertices which are labelled as 1, 2, and 3, say. Then from Equation (8.1) on page 35 of Tutte (1962) we have:

$$\psi_{n,0} \sim \frac{1}{16} \left(\frac{3}{2\pi} \right)^{\frac{1}{2}} n^{-\frac{5}{2}} \left(\frac{256}{27} \right)^{n+1}, \tag{11.7}$$

where n is the number of internal vertices (as $m = 0$) and the total number of vertices is $(n + 3)$.

Let M_n be the number of maximal planar subgraphs of K_n with labelled vertices. Then,

$$n!\psi_{n-3,0} = M_n 6(2n - 4), \qquad n \geq 4. \tag{11.8}$$

This is so because the left-hand side of (11.8) can be thought of as representing a Tutte triangulation with labelled vertices. Each triangulation has a face specified with vertices labelled 1, 2, and 3. Also there are $n!$ ways of labelling the n vertices. The right-hand side of (11.8) is the result of the following process. Take a maximal planar graph and embed it on the sphere. It is well known that this embedding is unique. Then find a region out of the $(2n - 4)$ possible regions and let this be the rooted region. There are six ways to label vertices 1, 2, and 3. Now (11.8) implies that

$$M_n = \frac{n!\psi_{n-3,0}}{12(n - 2)}. \tag{11.9}$$

Using (11.6), this reduces to

$$M_n = \frac{n(n - 1)(4n - 11)!}{6(n - 2)(3n - 7)!}. \tag{11.10}$$

Values for the first few M_n are given in Table 11.1. We can check the first few values quite easily. $M_4 = 1$ as the only solution is K_4 itself. For $n = 5$, each solution can be obtained by removing exactly one edge from K_5. As K_5 has 10 edges, $M_5 = 10$. For $n = 6$, a solution can be obtained by removing from K_6 either: (i) three non-adjacent edges or (ii) a three-edge path. There are 15 choices for (i) and (180) choices for (ii). Hence $M_6 = 195$.

n	M_n
4	1
5	10
6	195
7	5712
8	223440
9	10929600
10	641277000
11	4385962800
12	3424685806080
13	300495408594800
14	$2.9262949937020800 \times 10^{16}$
15	$3.1311876139568640 \times 10^{18}$
16	$3.6511299673744896 \times 10^{20}$
17	$4.6075561988281234 \times 10^{22}$
18	$6.2558644083184127 \times 10^{24}$
19	$9.0924582547729753 \times 10^{26}$
20	$1.4084536998888157 \times 10^{29}$
21	$2.3162980517300182 \times 10^{31}$
22	$4.0304466049683629 \times 10^{33}$
23	$7.3976935930983006 \times 10^{35}$
24	$1.4283603644534211 \times 10^{38}$

Table 11.1 Tabulation of M_n.

11.5 Puzzles and Games

In Sections 1.1.1. and 4.2, the Konigsberg bridge puzzle was introduced and analysed via Eulerian graphs. In Section 1.1.2 and 4.3, Hamilton's game was introduced via Hamiltonian graphs. The use of digraphs in game theory will be discussed in Section 12.4.1. In this section, we confine ourselves to graph theoretic devices which aid the analysis of various simple puzzles and board games.

11.5.1 Puzzles

A coloured cubes puzzle

We now describe a puzzle which can be solved elegantly by the rather nonintuitive construction of an appropriate edge-labelled pseudograph. It involves a given set of four cubes: C_1, C_2, C_3, and C_4, of identical dimensions. The faces of the cubes are painted with four given colours, denoted by R, G, Y, and B, so that each cube has at least one face of each colour.

A stack of the cubes, one on top of the other, with the stack having four parallel sides is to be formed. The puzzle involves finding, if possible, a stack so that the four faces on each of its lateral sides has distinct colours. For different assignments of the colours to the faces of the cubes, the puzzle may have none, one, or many different stacks representing solutions.

We now introduce a graph theoretic model of the puzzle which leads to an easy settlement of any instance of the puzzle. Construct a graph $G = (V, E)$, in which $V = \{R, G, Y, B\}$. Each cube C_i, contributes three edges (each labelled i) to G. An edge uv, where $u, v \in V$, labelled i, will be a member of E if and only if C_i has two opposite faces coloured u and v.

As an example, the cubes shown in Figure 11.4(a) have the associated colour structure graph shown in Figure 11.4(b). It is possible for any vertex to have a number of loops, indicating pairs of opposite faces coloured the same. Because each cube must possess each colour, each vertex must be incident with at least one edge labelled with each of the four edge labels: 1, 2, 3, and 4.

We now use the graph to search for a solution to the puzzle. Consider two opposite lateral sides of any stack representing a solution, if one exists. Each colour will appear twice, once on each side. The eight faces involved are represented by a spanning subgraph of the colour structure graph with four edges, all with distinct labels, such that each vertex is of degree two. The other two opposite lateral sides of the solution are represented by another subgraph, with the same properties, edge disjoint from the first. Further, any pair of subgraphs with these properties represents a solution. The two subgraphs in Figure 11.5 have the necessary properties and can be used to form a stack constituting a solution.

The farmer's puzzle

Consider a farmer standing on the bank of a river with a dog, a sack of wheat, and a goose. He wishes to transport these three and himself to the far bank by using a boat that will hold, at any one time, himself and at most one of his possessions. For obvious reasons the goose cannot be left alone with either the dog or the wheat. The puzzle is concerned with devising a sequence of crossings, in which the farmer progressively transports possessions across the river which terminates in all four being safely landed on the other bank.

We now devise a digraph model of this puzzle which can be analysed to find a solution. Let the dog, the wheat, and the goose be denoted by d, w, and g respectively. The state of the system, after any number of crossings, can be completely defined by which possessions are on the original bank after a round trip of two crossings, from and then back to, the original bank.

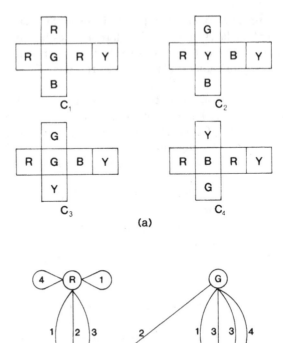

(a)

(b)

Figure 11.4 (a) Four coloured cubes and
(b) their colour structure graph

These states are:

$$\{d, w, g\}, \{d, w\}, \{d, g\}, \{w, g\}, \{d\}, \{w\}, \{g\}, \text{ and } \emptyset.$$

The states: $\{d\}$ and $\{w\}$ are infeasible because they constitute inadmissable combinations on the far bank. The objective is to find a sequence of crossings which transform the system from its initial state $\{d, w, g\}$, into the desired final state $\emptyset$, via a sequence of feasible states.

We can devise a digraph $D = (V, A)$, in which each feasible state is represented by a vertex in V, and uv is an arc in V if and only if it is possible to

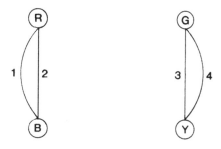

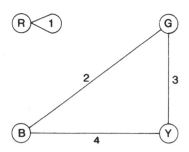

Figure 11.5 The two subgraphs constituting the
solution to the coloured cubes puzzle

transform the system from state u to state v by one round trip of crossings, beginning at the original bank, without creating an inadmissable combination of possessions on either bank. (Actually the very last system transformation comprises only a single crossing from the original bank to the far bank, and not the return crossing). The appropriate digraph is shown in Figure 11.6, where the infeasible states and their system transformations are not included.

Any solution to the puzzle can be represented by a path in the digraph from vertex dwg to vertex $\emptyset$. There are four such paths:

$$P_1 = \langle dwg, dw, dg, g, \emptyset \rangle, \qquad P_2 = \langle dwg, dw, wg, g, \emptyset \rangle,$$
$$P_3 = \langle dwg, dw, dg, wg, g, \emptyset \rangle, \qquad \text{and} \qquad P_4 = \langle dwg, dw, wg, dg, g, \emptyset \rangle.$$

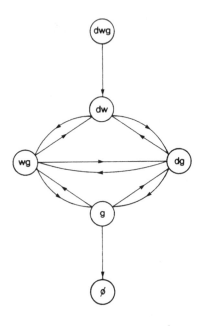

Figure 11.6 The digraph for the farmer's puzzle

P_1 and P_2 are more satisfactory in the sense that they involve less trips (three round trips and one one-way trip) than P_3 and P_4.

The solution corresponding to P_1 can be described as follows:

(i) Transport goose.

(ii) Return empty.

(iii) Transport wheat.

(iv) Return with goose.

(v) Transport dog.

(vi) Return empty.

(vii) Transport goose.

An acquaintance puzzle

Consider a set of six people in which it is clearly established whether or not each pair of people are acquainted or not. The puzzle involves discovering whether or not in the set there are either three mutual acquaintances or three mutual strangers. We can settle the puzzle by constructing an edge-coloured K_6 with a vertex for each of the six people. Each edge uv, in K_6 will be coloured *red* if the people represented by the vertices u and v are acquainted and coloured *blue* if they are strangers.

The subgraph of K_6 comprising the red (blue) edges is the *complement* of the subgraph comprising the blue (red) edges in the sense that these two spanning subgraphs are edge disjoint and their union of edges constitutes a complete graph.

The puzzle will be settled in the affirmative if it can be established that the edge-bicoloured K_6 has a monochrome triangle, i.e. all the edges of a 3-cycle are the same colour. This triangle is an example of a *clique*, a notion to be discussed further in the next section. To prove that such a triangle exists, consider any vertex, say v_1. It must be incident with at least three edges say: v_1v_2, v_1v_3, v_1v_4, of the same colour. Without loss of generality, assume that this colour is red. Now consider these three vertices: v_2, v_3, v_4, which are incident with these three red edges. If any two of v_2, v_3, and v_4 are joined by a red edge, then there exists a red triangle. Thus three people are all mutually acquainted. If not, then v_2, v_3, and v_4 induce a blue triangle, and there are three people: v_2, v_3, and v_4, who are all mutual strangers.

In terms of the above definitions, we have proven:

Theorem 11.1

If G is a graph with six vertices, then either G or its complement contains a triangle.

This puzzle leads to *Ramsey numbers*.

Definition

The *Ramsey number* $R_{p,q}$, is the smallest integer such that every graph with $R_{p,q}$ vertices contains either K_p or an independent set of q vertices.

Theorem 11.1 implies that $R_{p,q} \leq 6$. Indeed it can be shown that $R_{p,q} = 6$.

Chess board puzzles

Consider a regular chess board and the five chess pieces: rook (castle), bishop, knight, king, and queen. A collection of puzzles can be posed on the chessboard in which a person is assumed to have a large number of pieces of one type and none of the others. One problem is to decide how to place the maximum number of pieces of the same kind on the board (no more than one to a square) so that no capturing can occur.

As an example, let us consider the rook puzzle. We wish to place as many rooks as possible on the chessboard so that no two attack each other. A rook moves orthogonally, that is, through any number of unoccupied squares parallel to an edge of the board. Thus once a rook is placed it will not be possible to place another on the same rank (row of horizontal squares) or file (column of vertical squares). Thus the problem reduces to finding a set of squares such that:

   (i) No two are in the same rank or file, and

   (ii) The number in the set is a maximum.

Since a chessboard has only eight ranks (and files) it is obvious that no more than eight rooks can be placed. Hence the eight squares comprising the leading diagonal constitute a set satisfying properties (i) and (ii). This is only one of many optimal solutions to the problem — the others can be arrived at by interchanging the ranks of various pairs of rooks in the leading diagonal solution.

This puzzle of the rooks is straightforward to solve. We shall see that the puzzles associated with the other pieces are more challenging. Combinatorial methods (e.g. Yaglom and Yaglom (1964)) can be used to discover how many solutions there are to a given problem. In the discussion that follows we present a method of solution based on graph theory. There is no suggestion that these methods are superior to other methods for solving the chessboard problems. Rather, the problems are used as a vehicle to introduce the reader to the interesting and important topic of a *clique* in a graph.

It is appropriate to begin with a brief history of these puzzles and their solutions. The classic queen's puzzle originated with C.F. Gauss in 1850, where eight was quickly found to be the answer. But for the next sixty years, mathematicians were mainly concerned with the *Eight Queens Problem,* that is, the problem of enumerating how many different solutions to the queen's puzzle exist. At first Gauss concluded there were 76, then changed his mind to 72 and finally arrived at 92, which is correct. Then G. Bennett (1910) concluded that there are only 12 distinctly different solutions to the queen's problem (i.e., solutions that could not be obtained one from another by rotation or reflection of the chessboard).

The *Eight Queens Problem* was eventually extended to *The n-Queens Problem,* that is, solving the queen's puzzle for the general $n \times n$ chessboard. This was mentioned by J. W. Glaisher (1874) , who attempted to solve it using determinants. The answer was always suspected to be n queens, but remained uncertain until a clear proof was provided by Hoffman, Loessi and Moore (1969).

Many other puzzles using the chess pieces have been devised. For example, to solve the *knights tour puzzle*, a knight must tour the board using only legal moves, landing on every square exactly once. Another puzzle is to find the minimum number of pieces of one particular kind such that every square is either occupied or attacked by at least one piece; and, in addition, to count the number of different solutions. D. Kraitchik (1960) discussed the first puzzle in the book, *Mathematical Recreations* and R. W. Robinson (1975) has analysed the second problem. We now apply graph theory to analyse the non-attacking puzzles introduced earlier.

As we have seen in the previous section, it is possible to model associations between a given set of people by a graph. Suppose each person at a party is represented by a vertex. We create a graph by specifying that an edge is incident with vertices u and v if and only if the people represented by u and v know each other. Unless all at the party are strangers to each other, there is at least one subset of the people present with the property that everyone in the subset knows everyone else in the subset. Such a group is commonly called a *clique,* and in the graph model is represented by a subset of the vertices (also called a *clique*) such that every vertex in the subset is adjacent to every other vertex in the subset. A clique with the property that it no longer remains a clique when any further vertex is added is called a *maximal clique.*

We can now use this graph-theoretic machinery to analyse the chessboard puzzles. The approach is the same for all chess pieces. We begin by taking an arbitrary piece P, and creating a graph G, with 64 vertices — one for each square on the board. We then examine each square s, in turn and find which other squares would be attacked if P was placed on s. Edges are drawn in G according to the following rule: If a piece P on square s can attack a piece P on square t, then an edge joins s and t. When this has been done for all 64 squares we have created a graph for the puzzle concerned with the given piece P. The puzzle now reduces to finding a set V', of vertices in G with the properties:

(i) No two vertices in V' are adjacent.

(ii) The number of vertices in V' is a maximum.

It will become evident that for any choice of piece P, there are many sets which satisfy (i) and (ii). That is, there are many solutions to each puzzle. Suppose we have found a set V' satisfying (i) and (ii). Then (i) implies that each pair of vertices in V' must be adjacent in $\overline{G}$, the complement of G (defined in Section 2.3). That is, V' must constitute a clique in $\overline{G}$. But (ii) implies that V' must be a clique that has the maximum number of vertices among all cliques in $\overline{G}$. Therefore the puzzle becomes one of finding the maximal clique with the most vertices in $\overline{G}$. The general problem of finding a maximal clique in a graph is NP-complete.

Thus we can solve the chessboard puzzle by using a method to generate all the maximal cliques of $\overline{G}$ and then single out those having the maximum number of vertices. A maximal-clique-generating method has been presented by Bron and Kerbosch (1973) and we can use their technique as a subroutine in a computer program written to solve the puzzles. These graphs for a 3×3 board are shown in Figure 11.7, together with the corresponding puzzle solutions they produce.

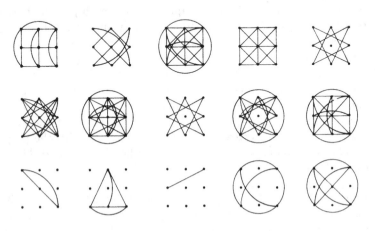

Figure 11.7 The 3x3 chessboard puzzles: top row, graphs; middle row, their complements; bottom row, maximal cliques and corresponding solutions (a) rooks, (b) bishops, (c) queens, (d) kings, and (e) knights

The method of Bron and Kerbosch is a DFS approach (see Section 9.5.2). In this case it explores various combinations of edges and cuts off lines of search which it establishes cannot lead to a clique. It generates cliques in a rather unpredictable order so as to try to minimize the number of search paths. It produces the large cliques first and it generates sequentially the cliques having a large common intersection. Sample solutions to the puzzle for 8×8 boards are displayed in Figure 11.8.

11.5.2 Games

Shannon's switching game

Consider the following game played between two players having alternate turns on an electrical network which has two distinct designated terminals,

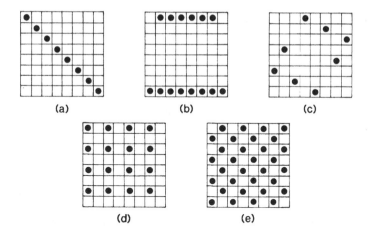

Figure 11.8 Sample solutions for the puzzles on an 8×8 board for (a) rooks, (b) bishops, (c) queens, (d) kings, and (e) knights

t_1 and t_2. At each turn, player A cuts the wire between any two junctions of the network, rendering it impossible for the wire to carry electrical current. At each turn player B protects the wire between any two junctions that has not yet been cut, preventing it from ever being cut by A.

The objective of A is to render it impossible for current to flow between t_1 and t_2, and the objective of B is to protect a set of wires which make it impossible for A's objective to be attained.

The game can be modelled by a graph G, in which the junctions (including t_1 and t_2) are represented by vertices in G and an edge joins vertices u and v in G, if and only if there is a wire directly connecting the corresponding junctions. At each turn A removes an edge of G. At each turn B protects an edge of G. Then A wins if the vertices in G representing t_1 and t_2 become disconnected and B wins if G eventually contains a path of protected edges joining t_1 and t_2. A sufficient condition for B to win the game is given in Theorem 11.2.

Theorem 11.2

B is guaranteed to win the game on a graph G, if and only if G contains two edge-disjoint spanning trees: (V, E_1) and (V, E_2), with $t_1, t_2 \in V$.

Proof

The reader is referred to the proof given by Busacker and Saaty (1965).

A square-filling game (Based on Section 5.5 of Robinson and Foulds (1980).)

Suppose there are two players, O and X, and they take turns (X first) to write their signs in one square in a row of five squares. The rule is that O and X may not be written on neighboring squares. The first person who cannot move loses.

Figure 11.9 shows the aborescence of all possible moves. We have simplified the situation somewhat by showing only one of any symmetrical pair of board positions. Having constructed the completed tree we label the sinks according to which player wins. We can then work back to the beginning to show that X must win if X always plays a move shown by the sequence A–B–E–K or A–B–E–L.

The argument is as follows: Consider a position such as that marked F. It is X's turn. If X fills the end square, O wins. So X chooses the middle square. But position F will never occur if O plays correctly, for in position C, O can ensure a win by choosing G rather than F. Applying this analysis to all positions, a perfectly played game will always run A–B–E–K or A–B–E–L, either resulting in a win for X.

11.6 Summary

We have given examples of the application of graph theory in areas outside the natural sciences and engineering. The social science applications outlined include organizational structure, social hierarchy, and consistency of choice. We then discussed a digraph model for the Leontief open input-output system in economics. We then went on to graph theoretic applications to stream networks in geography and the enumeration of floor plans in architecture. We ended the chapter with an excursion into graph theoretic approaches to various games and puzzles.

In this first chapter on the utility of graph theory, we have attempted to show that applications form an important part of the substance of the subject. We have pointed out various ways in which graph theory can be used, not only to specify problems rigorously, but also to solve them. We have limited ourselves to applications in the social sciences and related areas. These applications complement the deeper expositions in the natural sciences and engineering to come.

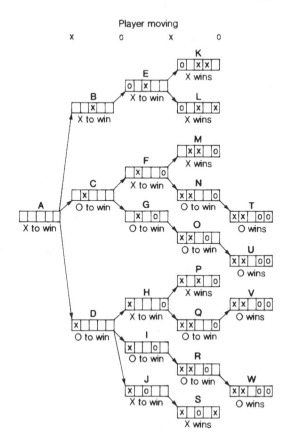

Figure 11.9 The tree of the square-filling game

11.7 Exercises

11.1 Construct a corporate structure chart for an organization of about six people, with which you are familiar. Draw a digraph representing the chart. Quantify the concept of *status* for individuals in the chart and calculate a measure of each person's status by finding shortest paths in the digraph.

11.2 Consider a group of rabbits : A, B, C, D, E, F, G, H, and I. The following matrices indicate parent-child relationships between them, in which a unit $i - j$ entry indicates that rabbit i is a child of rabbit j.

$$
\begin{array}{cc}
 & A \ \ B \\
\begin{array}{c} C \\ D \\ E \end{array} &
\left[\begin{array}{cc}
1 & 0 \\
1 & 1 \\
0 & 1
\end{array} \right]
\end{array}
\qquad
\begin{array}{cc}
 & C \ \ D \ \ E \\
\begin{array}{c} F \\ G \\ H \\ I \end{array} &
\left[\begin{array}{ccc}
1 & 0 & 0 \\
0 & 1 & 1 \\
0 & 1 & 1 \\
1 & 1 & 1
\end{array} \right]
\end{array}
$$

Draw appropriate digraphs which reflect the above relationships. Form appropriate matrices, which when multiplied together, represent the grandparent-grandchild relationships.   Draw a digraph representing these relationships and deduce the relationship between the matrix multiplication and the digraph construction process.

11.3 Suppose that a dog is given every possible pair of foods, one pair per day, from the set of foods: $F_1, F_2, \ldots, F_6$. The first preference of the dog within the pair is noted each day. Suppose that the dog chooses F_i first $(i - 1)$ times.

Draw the appropriate digraph representing these preferences. Is the dog consistent? When the dog was confronted with the F_3–F_2 pair, which food did it prefer?

11.4 Consider an open input-output Leontif model with:

$$
C = \begin{bmatrix}
0 & 2 & 7 & 1 \\
0 & 0 & 17 & 1 \\
0 & 10 & 0 & 2 \\
0 & 0 & 6 & 0
\end{bmatrix}
$$

Perform a complete analysis of the type explained in Section 11.2.

11.5 Find the orders, bifurcation ratios, and magnitudes of the stream network with adjacency matrix:

$$
\begin{bmatrix}
0 & 1 & 1 & 1 & 1 & 0 & 0 & 0 & 0 & 0 \\
1 & 0 & 0 & 0 & 0 & 0 & 0 & 0 & 0 & 0 \\
1 & 0 & 0 & 0 & 0 & 1 & 0 & 0 & 0 & 0 \\
1 & 0 & 0 & 0 & 0 & 0 & 1 & 1 & 0 & 0 \\
1 & 0 & 0 & 0 & 0 & 0 & 0 & 0 & 1 & 1 \\
0 & 0 & 1 & 0 & 0 & 0 & 0 & 0 & 0 & 0 \\
0 & 0 & 0 & 1 & 0 & 0 & 0 & 0 & 0 & 0 \\
0 & 0 & 0 & 1 & 0 & 0 & 0 & 0 & 0 & 0 \\
0 & 0 & 0 & 0 & 1 & 0 & 0 & 0 & 0 & 0 \\
0 & 0 & 0 & 0 & 1 & 0 & 0 & 0 & 0 & 0
\end{bmatrix}
$$

Construct two other 10-node stream networks, one with a higher, and one with a lesser probability of flooding.

11.6*Show that the number of *trivalent* (all vertices of degree three) planar graphs with a rooted region which have $n+1$ regions is $[2^{n-1}3!(3n-4)!]/(n-2)!2n!$. Hint: See Tutte (1962).

11.7 Show, using a graph theoretic approach, that the coloured cubes puzzle (Section 11.5.1) does not have a solution if, on each cube, there are three faces meeting at a corner which are blue (or any other single colour).

11.8 Solve the following problem using a graph theoretic approach. Three prisoners and three guards arrive at the bank of a river which they must cross. The only means of transportation is a two-person row-boat. All of the guards and one of the prisoners are capable of rowing. Devise a sequence of crossings which enable all six people to cross the river with the proviso that the prisoners never outnumber the guards on a bank if a guard is present on that bank.

11.9*Prove, using graph theoretic means, that if nine people are at a party, then either (i) four people are all mutual acquaintances or three people are all mutual strangers, or (ii) three people are all mutual acquaintances or four people are all mutual strangers. [This is equivalent to proving that the Ramsey number $R_{3,4}$ is 9.]

11.10*Prove all of the theorems stated in the chapter.

11.11 For each of the chess pieces: Queen, King, Rook, Knight, and Bishop, determine, using graph theoretic means, the minimum number of pieces of the one type of piece selected that can be placed on an $n \times n$ chessboard so that every square is either occupied or under threat by at least one piece.

12 Operations Research

Rather give me commentators plain, who with no deep researches vex the brain.

The Parish Register, George Crabbe

One of the common themes in operations research (OR) is the modelling approach. Unfortunately many accurate models of OR problems are intractable in the sense defined in Section 9.3. Of course a manager responsible for the efficient operation of some system will find no comfort in an OR consultant announcing that there is no known algorithm for the problem at hand which will provide an optimal solution in reasonable computional time. Come what may, the system has to be managed in the best way possible. If efficient algorithms are unavailable, there are a number of options open. The manager and the OR consultant can attempt to:

i) Develop methodology that will provide optimal solutions efficiently,

ii) Identify algorithms that will solve certain special cases of interest for the problem,

iii) Search for efficient algorithms that solve a relaxed version of the problem,

iv) Construct algorithms that run in reasonable computational time for most, but not all, problem instances, or

v) Give up the quest for optimality and provide approximate (or what are known as *heuristic*) methods which operate in reasonable computational time for all problem instances, but which have no guarantee of optimality.

One of the purposes of this chapter is to introduce some models from the mathematics of discrete OR. We shall show how useful insights for the analysis and solution of these models can be gained by viewing them in graph theoretic terms, when we wish to attain at least one of the five above aims. A knowledge of the concepts of Chapter 9 on algorithms, especially of Section 9.3 on complexity, is a useful background to the understanding of the present chapter.

We begin with a survey of various OR models and graph theoretic concepts or techniques which can be profitably used to analyse them. Our treatment is not encyclopaedic. We have deliberately chosen just a few examples of the major applications in order to give a flavour of this area. We then present some graph theoretic optimization algorithms which are useful for certain OR models. Finally we discuss transportation networks.

12.1 Operations research and graph theory

One simple use of graph theory in many fields, including OR, is the following. It is often convenient to depict the relationships between pairs of elements of a system by means of a graph or a digraph. Here the vertices of the graph represent the system elements and its edges or arcs (often weighted) represent the relationships between the elements. This approach is especially useful for transportation, scheduling, sequencing, allocation, assignment, and other problems which can be modelled as networks. Such a graph theoretic model is often useful as an aid in communicating or viewing a particular OR model. However this chapter is concerned not just with ways of representing OR problems, but also with using certain graph theoretic results to analyse or actually solve the problem. We now survey a few OR models which are amenable to solution by graph theoretic algorithms.

12.2 Graph theoretic algorithms in OR

In this chapter we adopt the following specialized definition of an algorithm. An *algorithm* for a problem is a scientific procedure which is guaranteed to converge to an optimal solution of the problem. This subsection discusses polynomial graph theoretic algorithms for certain OR problems.

12.2.1 Applications of Euler trails

Consider a mail carrier who begins delivering mail in a city suburb having picked it up from the post office. The carrier wishes to cover each street in the area at least once and finally return to the post office, travelling the least possible distance. This problem is known as the *Chinese Postman's Problem (CPP)*, as it was first considered by the Chinese mathematician Guan Mei-gu (1962). Further applications abound including: the collection of household refuse, milk delivery, the inspection of power, telephone, or railway lines, the spraying of salt-grit on roads in winter, office block cleaning, security guard routing, the scheduling of snow ploughs, and even museum touring.

The problem can be formulated as a graph theoretic model in which a weighted graph G is devised with its vertices, edges, and weights representing intersections, streets, and distances respectively. Recall from Section 4.2 that an *Euler trail* in a graph G, is a cycle in G which includes each edge exactly once. Clearly the CPP is equivalent to finding the least-weight Euler trail in an appropriate graph. If G is Eulerian any Euler trail is optimal. If G is not Eulerian, some edges will have to be covered more than once. One approach is to add duplicated edges of minimal total weight so as to make G Eulerian and then use an algorithm for detecting an Euler trail in the resulting graph. One such polynomial algorithm has been devised by Fleury. (See Kaufmann (1967).) More efficient algorithms have been described by Nijenhaus and Wilf (1975).

Good (1946) provided the following application of a directed Euler trail in a digraph. The position of a rotating drum in a computer can be recognized by means of binary signals produced at a number of electrical contacts on the surface on the drum. The surface is divided into 2^m sections, each comprising either conducting or insulating material. If an electrical contact is adjacent to a conducting (insulating) section the signal 1(0) is sent.

The problem is to minimize the number, say k, of consecutively placed contacts so that each position of the drum gives a unique reading. It can be solved as follows. Any position corresponds to a k-digit binary number. It can be shown that m contacts are sufficient by reducing the problem to finding a directed Euler trail in a certain digraph.

12.2.2 A timetabling problem

A school has m teachers: $T_1, T_2, \ldots, T_m$, and n classes: $C_1, C_2, \ldots, C_n$. Suppose that teacher T_i is required to teach class C_j for P_{ij} periods, and that it is desired to schedule a complete timetable in the minimum possible number of periods. The problem can be formulated as a graph theoretic model in which a bipartite graph $G = (V, E)$ is devised with $V = V_1 \cup V_2$

where:

$$V_1 = \{T_1, T_2, \ldots, T_m\} \qquad \text{and} \qquad V_2 = \{C_1, C_2, \ldots, C_n\},$$

and vertices T_i and C_j are joined by P_{ij} edges. The timetabling problem is equivalent to colouring the edges in V, so that no two adjacent edges have the same colour, with as few colours as possible — each colour representing a distinct period. (See Section 8.2 for an introduction to the colouring of graphs.) It is assumed that, in any given period, each teacher can conduct at most one class. We also assume that each class is taught by exactly one teacher. This leads to the fact that a teaching schedule for any given period can be represented by a matching in G. (See Section 8.3 for an introduction to matchings.) Also each matching in G represents a feasible (at least in terms of our assumptions) assignment of teachers to classes for one period. The timetabling problem of concern here is thus equivalent to finding a partition of E into as few matchings as possible. Bondy and Murty (1976) have presented an polynomial algorithm for bipartite graph edge colouring. They also solve the more realistic problem of assuming that there are only q classrooms available. The complication of preassignments (i.e. conditions specifying the periods during which certain teachers and classes must meet) has been studied by Dempster (1971) and by de Werra (1970). Many general timetabling problems are intractable.

12.2.3 The connector problem

A railway network connecting a number of cities is to be set up with the objective of making it possible to travel by some path between every pair of cities. Given the cost of construction w_{ij}, of linking cities v_i and v_j, *the connector problem* involves designing the network with the minimum possible construction cost. Other applications include: electrical network design (see Exercise 13.2), natural gas reticulation, communication network construction, and city utilities cable layout. A less direct application appears in the form of a subroutine in some solution methods for the *Travelling Salesman Problem*, which is discussed in Section 12.3.1. The problem can be formulated as a graph theoretic model in which a weighted graph G, is devised with vertices, edges, and weights representing cities, feasible connections, and construction costs respectively. The problem is equivalent of finding a minimum weight spanning tree of G. There are two well-known polynomial algorithms for this problem. They are discussed in Section 12.5.1.

12.2.4 Shortest path problems

Consider once again the connector problem just discussed. Before actually constructing the railway network it is conceivable that a planner may wish to calculate certain shortest paths in the graph of potential connections. There are a multitude of applications of the problem of finding shortest paths in a weighted graph. Polynomial algorithms for finding shortest paths in weighted graphs in which all weights are nonnegative are discussed in Section 12.5.2.

12.2.5 The assignment problem

Consider a factory in which n workers: $W_1, W_2, \ldots, W_n$, are to be assigned in a one-to-one fashion to n machines: $M_1, M_2, \ldots, M_n$. Each worker has been tested on each machine and a table of standardized times providing information about relative worker abilities are available. The problem is to assign the workers to machines so as to minimize the total of the standardized times of the assignment. The problem can be formulated as a graph theoretic model in which a weighted, bipartite graph $G = (V, E)$ is devised with $V = V_1 \cup V_2$ where $V_1 = \{W_1, W_2, \ldots, W_n\}$ and $V_2 = \{M_1, M_2, \ldots, M_n\}$. The edge joining W_i and M_j is weighted with t_{ij}, the standardized time for worker W_i on machine M_j. The problem is equivalent to finding a minimum-weight perfect matching in G, as discussed in Section 8.3. A polynomial algorithm for this problem, called the *Hungarian method*, has been given by Kuhn (1955). This problem, and the problem of Section 12.2.4 are special cases of a network flow model called the *capacitated transshipment model*, which is discussed in Section 12.6.

12.2.6 The location of centres

Consider a network of roads whose vertices represent communities. There are a number of emergency centres, such as hospitals, police, or fire stations to be located on the network, not necessarily in the communities. The optimality criterion is often taken to be the minimization of the distance of the furthest community to a centre. There is a second, related problem. For a given critical distance, locate the smallest number of centres so that all communities lie within this critical distance from at least one centre. The problem can be formulated as a graph theoretic model in which a weighted graph G, is devised with vertices, edges, and weights representing communities, roads, and distances respectively. The problem is to identify a set of vertices, possibly inserted into the edges of G, which satisfy the above criteria. Christofides (1975) has called these new vertices *absolute p-centres* and has provided an efficient algorithm which can solve either of the two above-mentioned problems.

With regard to the location problem just discussed, it is often the case that nonemergency facilities, such as switching centres for telephone networks, substations in electric power networks, post offices, libraries, or goods depots, are to be located. With such scenarios it is usual to modify the optimality criterion from one of *minimax* to one of *minimum*. That is, one wishes to minimize the total cost of travel from each community to the nearest centre. (These travel costs are commonly some function of the importance of the community, based on the demand for the service of each centre. Problems of this type come under the general heading of *facility location on a network*. Unfortunately there are no reported polynomial graph theoretic algorithms for this problem. The reader is referred to Hanan and Kurtzberg (1972) for further details on what is available.

12.2.7 Communications network reliability

Consider a number of centres which are to be connected by communications links. A measure of the reliability of the system is the smallest number of links whose breakdown makes communication between every pair of centres impossible. Each potential link, joining centres i and j, has a construction cost c_{ij}. The system is to be designed so that, for a given number k, at least k links must break down before complete pairwise communication becomes impossible. The optimality criterion is the total construction cost, which is to be minimized. The problem can be formulated as a graph theoretic model in which a weighted graph G is devised with vertices, edges, and weights representing centres, potential links, and construction costs respectively. The problem is to determine a minimum weight k-connected spanning subgraph of G. For $k = 1$, this problem reduces to the connector problem already discussed in Section 12.2.3. Unfortunately there are no known polynomial algorithms for problems for which $k > 1$. However the special case in which every possible link is available, and their construction costs are all equal, can be solved by the polynomial graph theoretic algorithm of Bondy and Murty (1976).

12.2.8 Project selection

Suppose an organization has n projects: $P_1, P_2, \ldots, P_n$ which must be carried out. Project P_i requires some subset $R_i \subset \{T_1, T_2, \ldots, T_q\}$, of the total set of q resources available. Each project can be completed in a single unit of time but projects requiring the same resource cannot be executed simultaneously. What is the maximum number of projects that could be executed at the same time? The problem may have to be solved repeatedly as more projects become available in later periods. It can be formulated as a graph theoretic model in which a graph G is devised with vertices and edges representing projects and resource nonoverlap requirement respectively. That is, P_i and P_j are connected whenever $R_i \cap R_j = \emptyset$. The problem reduces to finding a maximal independent set of G. A polynomial graph theoretic algorithm for this problem has been presented by Bron and Kerbosh (1973).

12.3 Graph theoretic heuristics in OR

Analysts in business and industry are often faced with problems of such complexity that the standard algorithms of OR are inappropriate. There are a number of reasons why this might be so: (1) The dimensions of the problems may be so large that the application of the fastest-known algorithm on the fastest computer may take a prohibitive amount of computational time. This is certainly true for certain vehicle routing problems. (2) The problems may be virtually impossible to formulate in explicit terms. The aims of different managers involved in operating a system may be conflicting or ill-defined. In fact, it may be difficult to express many features of a problem in quantitative terms. (3) Data collection may be beset with problems of accuracy and magnitude. For example, in large-scale location problems, the analyst may be faced with calculating an enormous number of location-to-location distances. In order to provide this information in reasonable time it may be necessary to make approximations. Sometimes the use of approximate data makes the concept of an optimal solution meaningless.

The idea of approximate methods, which are easy to use, but which give up the guarantee of optimality is not new. Indeed as early as 300 AD, Pappas, writing on Euclid, suggested this approach. R. Descartes and G.W. Leibnitz both attempted to formalize the subject. It became known as the study of *heuristic* whose aim was to investigate the methods of discovery and invention. It was allied with logic, philosophy, and psychology. The name itself was derived from the Greek word *heuriskein* — to discover. In OR today the term *heuristic* is used to describe a method which, on the basis of experience and judgement, seems likely to yield a good solution to a problem, but which cannot be guaranteed to produce an optimum. We now discuss some graph theoretic heuristics for various OR models.

12.3.1 The travelling salesman problem

The following scheduling problem arises in the pharmaceutical industry. Batches of n drugs: $D_1, D_2, \ldots, D_n$ are manufactured in a single reaction vessel, one at a time. If D_j is to follow D_i the vessel has to be cleaned, at a cost of c_{ij}. The batches are to be manufactured in a continuous, cyclic manner, so that once the last batch has been produced, the first batch is to be begun again. The problem is to find the production sequence with the least total cleaning cost. The problem can be formulated as a graph theoretic model in which a weighted graph G is devised with vertices, edges, and weights representing drugs, direct succession in the production sequence, and cleaning cost respectively. Then the problem is to determine a *Hamiltonian cycle* of G, of least cost. Recall from Section 4.3 that a Hamiltonian cycle in a graph G is a cycle which passes through each vertex of G exactly once. There are numerous other applications of this model including: mail box collection, school bus scheduling, electricity supply network design, and service vehicle routing. Unfortunately the problem is NP-hard (See Section 9.3, for an explaination of this term), and hence no efficient algorithms are known. However there are a number of effective heuristics for the problem including those by Lin (1965) and Christofides (1976), the latter guaranteeing a solution value within 50% of the optimum. A comprehensive treatise on the the travelling salesman problem has been compiled by Lawler et al. (1985).

12.3.2 A storage problem

Consider a factory which manufactures a number of chemicals which it then stores in a warehouse. Certain pairs of chemicals cannot be stored in the same compartment. What is the least number of compartments into which the warehouse must be partitioned for safe storage? This problem can be formulated as a graph theoretic model in which a graph G is devised with vertices and edges representing chemicals and incompatibilities respectively. The problem reduces to finding a minimum colouring of G. (See Section 8.2 for an introduction to graph colouring.) The model also has applications in bin packing, examination timetable construction, and resource allocation. Unfortunately the problem is NP-hard and thus a heuristic procedure is in order. Various available colouring heuristics have been compared by Matula et al. (1972), and Williams (1968) .

12.3.3 Mine ventilator shaft location

Consider an underground mining network of tunnels in which a ventilation system is to be located. The system comprises a pump and a number of ventilation units. The pump is usually sited at the main descent shaft. The units are usually located at the cutting faces (at the ends of the tunnels) although in an extensive network, additional units may be sited at internal locations as well. Each unit must be connected by some path of pipes, sometimes via intermediate units, to the pump. The cost of laying a pipe between any two locations in the network is known. The problem is to determine how to connect all the units to the pump with minimal laying cost. This problem can be formulated as a graph theoretic model in which a weighted graph $G = (V, E)$ is devised with vertices representing all possible sites and tunnel intersections, and edges representing all feasible connections between these locations, and weights representing laying costs. A subset $U \subseteq V$, is distinguished, representing the given set of locations of the pump and the units (the pump is not specially distinguished). The problem reduces to finding a subgraph of G of least cost which spans U. That is, it contains a path between every pair of vertices in U. The problem reduces to the *Steiner Problem in Graphs (SPG)*, which was discussed in Section 9.3. There we mentioned that it is NP-hard. Takahashi and Matsuyama (1980) have devised an efficient heuristic for the problem. Further results on the SPG are discussed in Section 9.3. A variation of the SPG is discussed in some depth in Section 15.3.

12.4 Digraphs in OR

Many practical systems which OR analysts study can be modelled as binary relations between their elements. Since this is what a digraph represents, digraph theory can often be applied to analyse OR problems, including those in Markov processes (see Rosenblatt (1957) and Howard (1971)), game theory, job sequencing, assembly line balancing, transportation, and resource planning. In this section we discuss effective digraph models and methods for certain OR problems.

12.4.1 Game theory

Game theory has developed into an important OR tool over the last 50 years as it can be used to find the best way to perform a set of tasks in a competitive environment. The *games* involved can be classified according to: the number of *players*, whether there is a stochastic element, whether complete information on the position of the game is available at every move, and whether there is a finite number of moves available at each position. The theory of digraphs is of limited use in the more general cases. However it can provide a useful approach, if not a complete analysis, for two-person, perfect information, deterministic, finite games. Such a game can be modelled by a *game digraph*, with vertices and arcs representing positions and legal moves respectively. A game digraph has a unique *starting vertex* representing the initial position and at least one *closing vertex* representing a position at which the game is terminated. The closing vertices are classified according to each outcome they represent, such as *a win for player A*, *a draw (tie)*, and *a win for player B*. Most game digraphs are acyclic, but when cycles are present there are rules to prevent endless cycle traversal. Each player tries to find a directed path from the starting vertex to a closing vertex which represents a win for that player. What each player is concerned with is, given the game has reached a certain vertex, can that player force a win? If either player can claim this, the vertex (position) is said to be *won*, as it is assumed that each player makes the best possible move at each vertex. The concept of the *kernel* of a digraph is useful in finding a winning strategy. A set of vertices K, in a digraph is termed a *kernel* if: (1) no two vertices in K are joined by an arc, and (2) every vertex v, not in K is joined by an arc (v, k) to some vertex $k \in K$. The following theorem is of relevance.

Theorem 12.1

Every acyclic digraph has a unique kernel.

Proof

Exercise

Assume that the player who makes the first move is labelled A, the other player, B. Assume further, that the player who makes the last move is the winner. Then we have an important result.

Theorem 12.2

If the starting point is not in the kernel of the game digraph, player A is assured of a win and A can win by always selecting vertices in K.

Proof

Exercise

Corollary 12.1

If the starting vertex is in the kernel, B is assured of a win by always selecting vertices not in K.

For further reading on this topic see Berge (1962) and Kaufmann (1967).

12.5 Optimization algorithms

In this section we present some graph theoretic algorithms which can be used to solve many OR models, including some discussed in the previous sections of this chapter.

12.5.1 Minimum spanning trees

The concept of a spanning tree was introduced in Section 3.6. The connector problem, requiring a spanning tree of minimum total weight (as a sum of the weights of its edges) in a weighted graph, was discussed in Section 12.2.3. Such a tree can be constructed by Kruskal's (1956) or Prim's (1957) algorithms. We begin with the former.

Kruskal's algorithm

Objective

To find a minimum weight spanning tree T, of a weighted graph $G = (V, E)$, with n vertices and edge weight w_{ij}, for each edge $ij \in E$.

Steps

(1) Sort the edges of E in order of nondecreasing edge weight.

(2) Examine each edge of E in the order created in (1). Accept this edge as part of T unless it creates a cycle* with the edges already in T. Settle ties arbitrarily.

(3) Terminate when T has $n - 1$ edges.

*Edges which would create such a cycle are rejected because their inclusion would make it impossible to form a tree.

We now illustrate Kruskal's algorithm by implementing it on a graph with 10 vertices whose edges have weights given in Table 12.1.

	2	3	4	5	6	7	8	9	10
1	8	5	9	12	14	12	16	17	22
2		9	15	17	8	11	18	14	22
3			7	9	11	7	12	12	17
4				3	17	10	7	15	18
5					8	10	6	15	15
6						9	14	8	16
7							8	6	11
8								11	11
9									10

Table 12.1

We begin by scanning Table 12.1 and finding the edge of least weight. This joins vertex 4 to vertex 5 and has weight 3. This edge is then accepted as a part of the solution. The next lowest edge weight is then identified — it joins vertex 1 to vertex 3 with weight 5. It is accepted as part of the solution. In selecting the next largest edge we have a choice: $\{5, 8\}$ and $\{7, 9\}$, both of weight 6. In this case an arbitrary choice is made, say $\{5, 8\}$, and this edge is accepted as part of the solution. The next shortest edge is naturally $\{7,9\}$ and it is also accepted. At the next stage there are a number of choices: $\{3,4\}$, $\{3,7\}$, and $\{4,8\}$, all of weight 7. If the first of these is chosen arbitrarily it can be accepted, as well as the second. However, the third choice, $\{4,8\}$, creates a cycle: $\langle 4,5,8,4 \rangle$. We know that the minimum solution will not contain a cycle, so we reject $\{4,8\}$. At the next stage there are even more choices: $\{1,2\},\{2,6\},\{6,9\},\{5,6\},and\{7,8\}$, all of weight 8. If all of the first three are accepted, the cycle $\langle 1,2,6,9,7,3,1 \rangle$ will be created. Suppose that $\{1,2\}$ and $\{2,6\}$ are chosen arbitrarily, and $\{6,9\}$ is rejected; $\{5,6\}$ and $\{7,8\}$ are also rejected as they create cycles. Moving on to the next stage we see that all of the choices of weight 9: $\{1,4\},\{2,3\},\{3,5\}$ and $\{6,7\}$, create cycles. Hence they are all rejected. Finally, we accept edge $\{9,10\}$ of weight 10 and the tree is complete, and has total weight,

$$3 + 5 + 6 + 6 + 7 + 7 + 8 + 8 + 10 = 60.$$

The minimum spanning tree produced by the algorithm is shown in Figure 12.1.

Prim's algorithm

Objective

To find the minimum weight spanning tree T of a weighted graph $G = (V, E)$ with n vertices and edge weight w_{ij}, for each edge $ij \in E$.

Steps

(1) Define the set of vertices in a component of vertices to be built up to be the set C. Let

   $C = \{i,j\}$, where edge ij is the edge of least weight;

   $T = \{ij\}$, where T is the set of edges which are currently part of the final spanning tree.

(2) For all $i \in V \backslash C$, find a vertex $k \in C$ such that

   $$w_{ik_i} = \min_{j \in C}\{w_{ij}\} = \alpha_i.$$

   Set i to have a label $\{k_i, \alpha_i\}$. If no such label can be found, set i to have label $[0, \infty]$.

(3) Identify i^* such that

   $$\alpha_{i^*} = \min_{i \in V \backslash C}\{\alpha_i\}.$$

   Set

   $$C \leftarrow C \cup \{i^*\}, \quad T \leftarrow T \cup \{k_{i^*} i^*\}$$

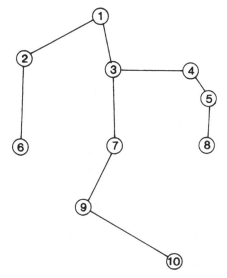

Figure 12.1 A minimum spanning tree
derived by Kruskal's method

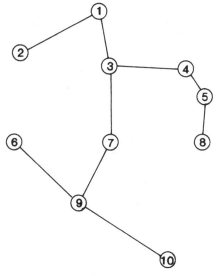

Figure 12.2 A minimum spanning tree
derived by Prim's method

If $|C| = n$, terminate. Otherwise, continue.

(4) For all $i \in V \backslash C$ such that i^*i is an edge in E and $\alpha_i > w_{i^*i}$, set $\alpha_i = w_{i^*i}$ and $k_i = i^*$. Go back to step 3.

We explain Prim's method by using it to solve the example problem just discussed. As in Kruskal's method, we begin by accepting the edge of least weight which is $\{4,5\}$ of weight 3. The two vertices incident with this edge, 4 and 5, belong to *component* C. We now find the vertex not in C which is closest (requiring the least weight edge to connect) to a vertex in C. That is, we consider vertices 4 and 5 and find the vertex closest to them. Table 12.1 reveals that it is vertex 8, which is 6 from vertex 5. Edge $\{8,5\}$ is accepted as part of the solution. This creates component C with vertices 4, 5, and 8. We repeat this step and look for the closest vertex to one of these three. It is vertex 3 which is 7 from vertex 4. Edge $\{3,4\}$ is accepted. Note that, unlike Kruskal's method, edge $\{4,8\}$ is not considered because it links two vertices already in P. Vertex 3 is added to C which now contains vertices: 3, 4, 5, and 8.

Continuing this process we see that the next vertex to be included in C is vertex 1 which is 5 from vertex 3. Edge $\{1,3\}$ is accepted. Next vertex 7 is included as it is 7 from vertex 3. Edge $\{3,7\}$ is accepted. Then vertex 9, which is 6 from vertex 7, is included. Edge $\{7,9\}$ is accepted and C contains vertices 4, 5, 8, 3, 1, 7, and 9. At the next iteration we see that vertices 2 and 6 are both 8 from vertices 1 and 9, respectively. This is the first tie and it is settled arbitrarily. Suppose that edge $\{1,2\}$ is accepted and vertex 2 is included in C. Edge $\{6,9\}$ can also be accepted and vertex 6 is included. This leaves only vertex 10 not in C. It is closest to vertex 9 so that edge $\{9,10\}$ is accepted and C contains all the vertices. When this occurs the algorithm is terminated. It has a different structure from that derived by Kruskal's method in Figure 12.1, but of course, the same total weight:

$$3 + 6 + 7 + 5 + 7 + 6 + 8 + 8 + 10 = 60.$$

The minimum spanning tree produced by the algorithm is shown in Figure 12.2.

Algorithm effectiveness

The solutions generated by the two algorithms, as shown in Figure 12.1 and 12.2, are different. However, either solution could have been attained by the other algorithm by settling ties in a different manner. Indeed both algorithms guarantee to find all minimum solutions to every problem to which they are applied. However Prim's algorithm is, in general, more efficient than Kruskal's algorithm. Effective implementations of Prim's algorithm have been devised by Dijkstra (1959) and by Kevin and Whitney (1972).

12.5.2 Shortest paths

The concept of a path in a graph was introduced in Section 2.1. The shortest path problem in a weighted graph was discussed in Section 12.2.4. In this section we explain algorithms for finding shortest paths (as a sum of the weights of its edges) between one specified vertex and all others, and between all pairs of vertices in a weighted graph which does not contain cycles of negative weight. We explain Dijkstra's algorithm (1959) for the former problem and Floyd's algorithm (1962) for the latter. Algorithms for shortest paths in graphs which may possess negative edge weights, and hence cycles of negative weight, have been devised by Ford (1956), Moore (1957), Bellman (1958), and Dantzig (1967). An application of this last-mentioned algorithm is given in Section 14.2.5.

Dijkstra's algorithm

Objective

To find the shortest paths from one specified vertex, say p, in a weighted graph $G = (V, E)$ with weight w_{ij} for edge $ij \in E$, to all other vertices in V.

Steps

(1) Define $D(i)$ as the shortest distance from p to i yet known, for all $i \in V$. Set $D(p) = 0$. Mark this label of p as *permanent*. Set $D(i) = \infty$ for all $i \in V \backslash \{p\}$. Mark these labels as *temporary*. Let A equal the set of vertices i in V, for which $D(i)$ has been defined. That is, A is the set of vertices labelled so far with permanent labels. Let $B = V \backslash A$. Set $A = \{p\}$.

(2) If $B = \emptyset$ proceed to step (4). Otherwise continue. Update the label of each vertex j in B as

$$D(j) = \min_{i \in A}\{D(j), w_{ij} + D(i)\}.$$

Let k be such that
$$D(k) = \min_{j \in B}\{D(j)\}.$$

(3) $A \leftarrow A \cup \{k\}$ and $B \leftarrow B \backslash \{k\}$. Mark k with $D(k)$ as its permanent label. Go back to step (2).

(4) List all pairs i and $j \in V$ for which

$$D(j) = w_{ij} + D(i). \tag{12.1}$$

Each edge ij for which (12.1) holds is on a shortest path from p to j.

(5) Terminate.

The Dijkstra procedure can be terminated early if not all shortest paths from p are required. Once the labels for all the vertices in the given subset have been calculated, step (4) can be performed to find the actual paths.

The identification of the actual paths relies on the property that if the shortest path from p to j passes through i, then the p to i part of the p to j path constitutes a shortest path from p to i.

Steps (2) and (4) may create a number of ties. That is, there may be a number of different edges which all correspond to the minimum value sought. The method breaks ties arbitrarily. That is, the method makes a choice at random. The existence of ties implies that there are alternative shortest paths.

We illustrate Dijkstra's method by using it to find the shortest paths from vertex $p = 1$, to all other vertices in the graph shown in Figure 12.3 whose weight matrix is given in Table 12.2.

	2	3	4	5	6	7	8	9	10
1	80	50	90	∞	∞	∞	∞	∞	∞
2		90	∞	∞	80	110	∞	∞	∞
3			70	∞	∞	70	120	∞	∞
4				30	∞	∞	70	∞	∞
5					∞	∞	60	∞	∞
6						90	∞	80	∞
7							80	60	110
8								∞	110
9									100

Table 12.2

A label $D(i)$, is defined for each vertex i as the shortest distance from p to i. Naturally we set $D(1) = 0$, as $p = 1$. The labels for the other vertices are as yet undefined. We begin by finding the vertex which is closest to P. A scan of the first row in Table 12.2 reveals that it is vertex 3. Thus we can set $D(3) = 50$. The algorithm then calculates the other labels, one for each iteration. When all the vertices are labelled, it calculates the actual shortest paths which go with these shortest distances. The general step of the labelling process is as follows. Identify each edge which joins a labelled vertex to an unlabelled vertex . At this instant, as vertices 1 and 3 are labelled, such edges are:$\{1,2\}, \{1,4\}, \{3,2\}, \{3,4\}, \{3,7\}$, and $\{3,8\}$. The reader will find it beneficial to make a copy of Figure 12.3 and label the vertices as indicated. We now calculate the various possibilities for shortest

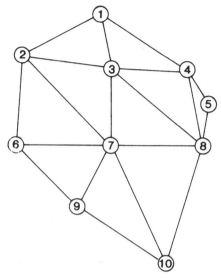

Figure 12.3 The graph whose weight matrix is
given in Table 12.3

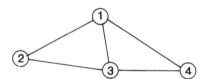

Figure 12.4 The graph whose weight matrix
is given in Table 12.4

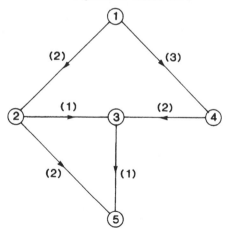

Figure 12.5 A network with source 1 and sink 5

weights involved with these edges. The $\{1,2\}$ edge implies that we could travel from vertex 1 to vertex 2 in $D(1)+c_{12} = 0+80 = 80$ units. Similarly, $D(1) + c_{14} = 0 + 90 = 90$ units.

From the first calculation, we know that we can travel to vertex 2 in no more than 80 units. Further, $D(3) + c_{32} = 50 + 90 = 140$, $D(3) + c_{34} = 50+70 = 120$, $D(3)+c_{37} = 50+70 = 120$, and $D(3)+c_{38} = 50+120 = 170$.

The smallest weight produced by these calculations is the first and $D(2)$ is set to 80. The general step is repeated and the reader should check that it leads to setting $D(4) = 90$. In successive iterations the following labellings occur: $D(5) = 120$, $D(7) = 120$, $D(6) = 160$, $D(8) = 180$, $D(9) = 180$ and $D(10) = 230$.

Having labeled all of the vertices, a backtracking procedure is invoked to find the actual shortest paths themselves. We know that the minimum weight required to travel from vertex 1 to vertex 10 is 230 units. This value of $D(10) = 230$ was arrived at by adding to the label of a neighbouring vertex, the weight of the appropriate edge. That is, the path from 1 to 10 of 230 travels through exactly one of 7, 8, and 9. Which of the following is true:

$$D(9) + c_{9,10} = D(10)?$$
$$D(8) + c_{8,10} = D(10)?$$
$$D(7) + c_{7,10} = D(10)?$$

As the last equation is the only one which is true, the shortest path to 10 must pass through 7. Repeating this process for vertex 7, we find that $D(3)+c_{37} = D(7)$. Hence the shortest path passes though vertex 3 as well. The complete 1–10 path is $\langle 1 - 3 - 7 - 10 \rangle$ with a weight of $D(10) = 230$. The complete list of shortest paths is given in Table 12.3.

Vertex	Path	Weight		
1	1	$D(1)$	=	0
2	1–2	$D(2)$	=	80
3	1–3	$D(3)$	=	50
4	1–4	$D(4)$	=	90
5	1–4–5	$D(5)$	=	120
6	1–2–6	$D(6)$	=	160
7	1–3–7	$D(7)$	=	120
8	1–3–8	$D(8)$	=	170
9	1–2–7–9	$D(9)$	=	180
10	1–3–7–10	$D(10)$	=	230

Table 12.3

Floyd's Algorithm

Objective

To find the shortest path between each pair of vertices in a weighted graph $G = (V, E)$ with n vertices and weight w_{ij} for edge $ij \in E$.

Steps

(1) Define matrix $D = (d_{ij})_{n \times n}$, where d_{ij} is the length of the shortest path from vertex i to vertex j. Initially, set $d_{ij} = w_{ij}$ for all i and j. Define matrix $P = (p_{ij})$, where p_{ij} is the second vertex on the shortest path from vertex i to vertex j. Initially, set $p_{ij} = j$ for all j.

(2) For all k, $k = 1, 2, \ldots, n$, and for all i, $i = 1, 2, \ldots, n$ and for all j, $j = 1, 2, \ldots, n$, perform steps (3) and (4).

(3) Set d_{ij} to become the minimum of d_{ij} and $d_{ik} + d_{kj}$.

(4) Whenever $d_{ij} > d_{ik}$ in step (3), set p_{ij} to become p_{ik}.

We illustrate Floyd's method by using it to find the shortest path between each pair of vertices in the graph shown in Figure 12.4, whose weight matrix is given in Table 12.4.

	1	2	3	4
1	0	48	30	54
2	42	0	40	∞
3	36	60	0	91
4	48	∞	48	0

Table 12.4

Dijkstra's method, explained in the last section, could be used repeatedly with each vertex successively replacing p. This is an inefficient way to tackle the problem as we are using the method in a way for which it was not designed. This leads to each shortest path being calculated twice — once in each direction. Floyd's method is a better way to go about the task.

In order to implement Floyd's algorithm, five 4×4 matrices: D^k, $k = 0$, 1, 2, 3, 4, are defined. Note that k is not a power but a superscript. The entry d_{ij}^k of D^k, is defined to be the length of the shortest path from vertex i to vertex j with intermediate vertices on the path chosen from the first k vertices: $1, 2, \ldots, k$. Naturally D^0 is just Table 12.4, as these are the direct distances with no intermediate vertices on the paths. One of our aims is to calculate D^4 whose entries are the lengths of the shortest paths with any of the vertices as intermediates. The other aim is to find the paths themselves. For this we define another 4×4 matrix $P = (p_{ij})_{n \times n}$, in which p_{ij} is defined to be the second vertex on the shortest path from vertex i to vertex j. The entries in P are updated as the algorithm proceeds and they are not necessarily correct, according to the definition of p_{ij}, until the method is terminated.

The matrices D, are calculated in the order: D^0, D^1, D^2, D^3, D^4. We shall not know all the correct entries of P until D^4 is calculated. We begin by defining P as

$$P = \begin{bmatrix} 1 & 2 & 3 & 4 \\ 1 & 2 & 3 & 4 \\ 1 & 2 & 3 & 4 \\ 1 & 2 & 3 & 4 \end{bmatrix}.$$

The rationale for this is that we do not necessarily know initially any of the shortest paths. So to start the process we assume that each shortest path from vertex i to vertex j is simply the edge joining i to j. Thus the i-j entry in P is j, for each i-j pair, because it is defined to be the second vertex on the path.

The method begins by calculating D^1 from D^0 and updating P in the process. Recall that the i-j entry in D^1 contains the shortest weight from vertex i to vertex j, possibly via vertex 1. Naturally the first row (and the first column) of D^1 will be identical to the first row (and the first column) of D^0. The first entry that can possibly differ is d_{23}^1. Is it worthwhile to go through vertex 1 when travelling from vertex 2 to vertex 3? We have to compare $d_{23} = 40$ with $d_{21} + d_{13} = 42 + 30 = 72$. (The superscript 0 is dropped and d_{ij}^0 is written as d_{ij}.) We find that it is not worthwhile and $d_{23}^1 = d_{23} = 40$. We now calculate d_{24}^1. From Table 12.4, $d_{24} = \infty$, indicating that there is no edge from vertex 2 to vertex 4. However $d_{21} + d_{14} = 42 + 54 = 96$. Hence we set $d_{24}^1 = 96$. We must now update P. The second vertex on the shortest known path from vertex 2 to vertex 4 is no longer vertex 4 but vertex 1.

Thus p_{24} is updated to become 1. Calculating d_{32}^1, we find that $d_{32} = 60$ and $d_{31} + d_{12} = 36 + 48 = 84$, so $d_{32}^1 = 60$ and no change is made to P. Calculating d_{34}^1, we find that $d_{34} = 91$ and $d_{31} + d_{14} = 36 + 54 = 90$. As we can shave one unit off the length of the path from vertex 3 to vertex 4 (by passing through vertex 1), we set $d_{34}^1 = 90$ and set $p_{34} = 1$. The reader should verify that in calculating the fourth row of D^1, $d_{42}^1 = 96$, $d_{43}^1 = 48$, and p_{42} is updated to 1. Thus

$$D^1 = \begin{bmatrix} 0 & 48 & 30 & 54 \\ 42 & 0 & 40 & 96 \\ 36 & 60 & 0 & 90 \\ 48 & 96 & 48 & 0 \end{bmatrix},$$

and

$$P = \begin{bmatrix} 1 & 2 & 3 & 4 \\ 1 & 2 & 3 & 1 \\ 1 & 2 & 3 & 1 \\ 1 & 1 & 3 & 4 \end{bmatrix}.$$

We now calculate D^2 from D^1. Now d_{ij}^2 contains the shortest distance from vertex i to vertex j by possibly passing through *both* vertices 1 and 2. The first entry that can possibly be updated is d_{13}^2. Now $d_{13}^1 = 30$. But $d_{12}^1 + d_{23}^1 = 48 + 40 > 30$. Hence $d_{13}^2 = 30$.

Continuing in this way we set: $d_{14}^1 = 54$, $d_{12}^1 + d_{24}^1 = 48 + 96$, and $d_{14}^2 = 54$.

The second row (and second column) of D^2 must equal the appropriate entries in D^1 and so do not need to be checked:

$$\begin{aligned}
d_{31}^1 &= 36 & d_{32}^1 + d_{21}^1 &= 60 + 48, & \text{set} \quad d_{31}^2 &= 36; \\
d_{34}^1 &= 90, & d_{32}^1 + d_{24}^1 &= 60 + 96, & d_{34}^2 &= 90; \\
d_{41}^1 &= 48, & d_{42}^1 + d_{21}^1 &= 96 + 42, & d_{41}^2 &= 48; \\
d_{43}^1 &= 48, & d_{41}^1 + d_{13}^1 &= 48 + 30, & d_{43}^2 &= 48.
\end{aligned}$$

Thus

$$D^2 = \begin{bmatrix} 0 & 48 & 30 & 54 \\ 42 & 0 & 40 & 96 \\ 36 & 60 & 0 & 90 \\ 48 & 96 & 48 & 0 \end{bmatrix},$$

and

$$P = \begin{bmatrix} 1 & 2 & 3 & 4 \\ 1 & 2 & 3 & 1 \\ 1 & 2 & 3 & 1 \\ 1 & 1 & 3 & 4 \end{bmatrix}.$$

We now calculate D^3 from D^2.

$$\begin{aligned}
d_{12}^2 &= 48, & d_{13}^2 + d_{32}^2 &= 30 + 60, & \text{set} \quad d_{12}^3 &= 48; \\
d_{14}^2 &= 54, & d_{13}^2 + d_{34}^2 &= 30 + 40, & d_{14}^3 &= 54; \\
d_{21}^2 &= 42, & d_{23}^2 + d_{31}^2 &= 40 + 36, & d_{21}^3 &= 42; \\
d_{24}^2 &= 96, & d_{23}^2 + d_{34}^2 &= 40 + 90, & d_{24}^3 &= 96; \\
d_{41}^2 &= 48, & d_{43}^2 + d_{31}^2 &= 48 + 36, & d_{41}^3 &= 48;
\end{aligned}$$

$$d_{42}^2 = 96, \qquad d_{43}^2 + d_{32}^2 = 48 + 60, \qquad d_{42}^3 = 96.$$

$$D^3 = \begin{bmatrix} 0 & 48 & 30 & 54 \\ 42 & 0 & 40 & 96 \\ 36 & 60 & 0 & 90 \\ 48 & 96 & 48 & 0 \end{bmatrix},$$

and

$$P = \begin{bmatrix} 1 & 2 & 3 & 4 \\ 1 & 2 & 3 & 1 \\ 1 & 2 & 3 & 1 \\ 1 & 1 & 3 & 4 \end{bmatrix}.$$

The final step is to calculate D^4 from D^3.

$$\begin{aligned}
d_{12}^3 &= 48, & d_{14}^3 + d_{42}^3 &= 54 + 96, & \text{set } d_{12}^4 &= 48; \\
d_{13}^3 &= 30, & d_{14}^3 + d_{43}^3 &= 54 + 48, & d_{13}^4 &= 30; \\
d_{21}^3 &= 42, & d_{24}^3 + d_{41}^3 &= 96 + 48, & d_{21}^4 &= 42; \\
d_{23}^3 &= 40, & d_{24}^3 + d_{43}^3 &= 96 + 48, & d_{23}^4 &= 40; \\
d_{31}^3 &= 36, & d_{34}^3 + d_{41}^3 &= 90 + 48, & d_{31}^4 &= 36; \\
d_{32}^3 &= 60, & d_{34}^3 + d_{42}^3 &= 90 + 96, & d_{32}^4 &= 60.
\end{aligned}$$

Thus

$$D^4 = \begin{bmatrix} 0 & 48 & 30 & 54 \\ 42 & 0 & 40 & 96 \\ 36 & 60 & 0 & 90 \\ 48 & 96 & 48 & 0 \end{bmatrix},$$

and

$$P = \begin{bmatrix} 1 & 2 & 3 & 4 \\ 1 & 2 & 3 & 1 \\ 1 & 2 & 3 & 1 \\ 1 & 1 & 3 & 4 \end{bmatrix}.$$

The times of the shortest paths are given in D^4 and the shortest paths themselves can be found by interpreting P. An an example, to find the shortest path from vertex 3 to vertex 4 we see that $p_{34} = 1$. Thus the path begins as $\langle 3 - 1 \rangle$. We now calculate p_{14}, the second vertex on the shortest path from vertex 1 to vertex 4. As $p_{14} = 4$, the complete path is $\langle 3 - 1 - 4 \rangle$. The other paths can be found in a similar manner.

12.5.3 An Introduction To Transportation Networks

The concept of a network was introduced in Section 1.2. Consider a network $N = (V, A)$, with n vertices, source v_1, sink v_n, along the arcs of which a commodity is to flow from v_1 to v_n. It is assumed that there is an upper capacity u_{ij}, on the ability of each arc $v_i v_j$ in A to accommodate the flow of this capacity and, in some scenarios, a unit cost c_{ij} for flow in $v_i v_j$. It is further assumed that there is *conservation of flow* at each vertex in N, other than at v_1 and v_n. That is, the sum of the quantities flowing into each intermediate vertex v, is equal to the sum of the quantity flowing out of v, for all $v \in V$ except for v_1 and v_n.

Let x_{ij} be the quantity of the commodity flowing in from vertex v_i to vertex v_j, b be the total quantity leaving v_1 and entering v_n, and u_{ij} be the upper capacity of arc $v_i v_j$, to accommodate flow.

Then

$$\sum_{j=1}^{n} x_{1j} - \sum_{i=1}^{n} x_{i1} = b, \tag{12.2}$$

$$\sum_{i=1}^{n} x_{ik} - \sum_{j=1}^{n} x_{kj} = 0, \qquad \text{for all } k = 2, 3, \ldots, n-1, \tag{12.3}$$

$$\sum_{i=1}^{n} x_{in} - \sum_{j=1}^{n} x_{nj} = b, \text{and} \tag{12.4}$$

$$0 \le x_{ij} \le u_{ij}, \qquad \text{for all } v_i v_j \in A. \tag{12.5}$$

Constraint (12.2) represents the fact that b units of flow are to leave the source v_1. Constraints (12.3) represents the conservation of flow at intermediate vertices. Constraint (12.4) represents the fact that b units of flow must arrive at the sink v_n. Constraints (12.5) represents the capacity constraints on the arcs. Networks with the above structure are called *transportation networks*.

If a *return arc* $v_n v_1$, is added to the network, and is assigned flow $x_{n1} = b$, *circulation flows* result, with conservation of flow at *all* vertices. Also, the zero in (12.5) can be replaced by a strictly positive bound. Simple algorithms for some basic transportation network problems will be explained now and more advanced techniques will be mentioned in Section 12.6. An application of a transportation problem is given in Section 16.2. We begin by formulating a simple transportation network problem and providing a straight-forward algorithm for it.

The Maximum Flow Problem

We now define the *maximum flow problem* for a given transportation network $N = (V', A)$ with n nodes. It is required to assign flow to the arcs of A so that the maximum quantity possible, subject to arc capacity and the conservation of flow, leaves the source v_1 of N, and arrives at the sink v_n. That is, the problem is to

$$\text{Maximize } b \qquad\qquad (12.6)$$

subject to (12.2), (12.3), (12.4), and (12.5).

Definitions

A cut-set in a transportation network which partitions the source and the sink into different vertex subsets is called a *cut*.

The *capacity* of a cut in a transportation network is equal to the sum of the capacities of its arcs.

The cut of least capacity in a transportation network is termed the *minimal cut*.

The following theorem is helpful in devising an algorithm for this problem.

Theorem 12.3 (The Max-Flow, Min-Cut Theorem).

In any transportation network N, the maximum source-to-sink flow possible is equal to the capacity of the minimal cut in N.

Proof

Delayed until the end of this section.

We use Theorem 12.3 to motivate an algorithm for the maximum flow problem by considering an instance of the problem given by the network in Figure 12.5, with source vertex 1, sink vertex 5, and arc capacities as shown. The problem is to assign flow to the arcs, subject to (12.2) – (12.5) so as to maximize the source to sink flow.

The total amount of flow leaving vertex 1 is

$$x_{12} + x_{14}.$$

According to our definition, we can set this equal to b.

$$x_{12} + x_{14} = b.$$

This total quantity b, must ultimately reach vertex 5 as we have assumed conservation of flow. The amount of flow reaching vertex 5 is

$$x_{25} + x_{35}.$$

Thus we can set

$$x_{25} + x_{35} = b.$$

We also have a constraint for each intermediate vertex based on the conservation of flow. Consider vertex 2. The amount of flow entering vertex 2 is x_{12}. The amount of flow leaving vertex 2 is

$$x_{23} + x_{25}.$$

Thus

$$x_{12} = x_{23} + x_{25},$$

or

$$x_{12} - x_{23} - x_{25} = 0.$$

We can build up a similar constraint for each other intermediate vertex. The last set of constraints is based on the capacity of each arc. We also assume that the amount of flow that each arc can accomodate is nonnegative. Thus we have

$$0 \le x_{ij} \le u_{ij}, \qquad \text{for each arc } (i, j).$$

The complete mathematical formulation is

$$\text{Maximize } b$$

subject to

$$x_{12} + x_{14} = b,$$
$$x_{12} - x_{23} - x_{25} = 0,$$
$$x_{23} + x_{43} - x_{35} = 0,$$
$$x_{14} - x_{43} = 0,$$
$$x_{25} + x_{35} = b,$$

$$0 \le x_{12} \le 2, \qquad 0 \le x_{23} \le 3, \qquad 0 \le x_{14} \le 3,$$
$$0 \le x_{43} \le 2, \qquad 0 \le x_{25} \le 2, \qquad 0 \le x_{35} \le 1.$$

Cuts and Labelling

On looking at Figure 12.5, we can see that if arcs (2,5) and (3,5) were removed, vertex 1 would be cut off from vertex 5. The capacity of the cut in question is thus,

$$c_{25} + c_{35} = 2 + 1 = 3.$$

Theorem 12.3 allows us to recognize when we have found the maximum possible flow. The labelling method, which we now explain, progressively loads the network with more source-to-sink flow. Once the total equals the capacity of the minimal cut, an optimal solution is at hand.

The labelling method

The reader may have noticed that we defined x_{ij} in Section 12.5.3 so that it was theoretically possible to have flow from vertex j to vertex i along arc (v_i, v_j) in the backwards direction. Although this is usually impossible in reality for technical reasons, we find the possibility a useful mathematical device. For each arc (v_i, v_j), we define an oppositely directed arc (v_j, v_i) with zero capacity. Flows in opposite directions in a single arc have their magnitudes subtracted, one from the other, to produce a single flow in the direction of the larger flow. To illustrate this, if we have $x_{14} = 3$ and $x_{41} = 2$, the result is a flow of 1 unit from v_1 to v_4. That is, an adjustment is made and $v_{14} = 1$ and $v_{41} = 0$. The *excess capacity* e_{ij}, of an arc (v_i, v_j), when it has flow x_{ij}, is defined to be the additional flow that it could accommodate up to its capacity u_{ij}. That is,

$$e_{ij} = u_{ij} - x_{ij}.$$

This concept allows us to change our minds about previous assignments and to modify them. For instance, suppose we have a flow $x_{23} = 0$ in arc (v_2, v_3) with capacity $u_{23} = 1$. Now $e_{23} = u_{23} - x_{23} = 1 - 0 = 1$. The imaginary arc (v_3, v_2) has flow $x_{32} = 0$ and capacity $u_{32} = 0$. Thus $e_{32} = 0$. Now suppose that we decide to assign a flow of $x_{23} = 1$ to (v_2, v_3). The excess capacity of (v_2, v_3) is reduced to $e_{23} = u_{23} - x_{23} = 1 - 1 = 0$. The excess capacity of arc (v_3, v_2) is increased by the amount just assigned; that is, it is set to $e_{32} = 1$. Although it is impossible in practice for (v_3, v_2) to accommodate any flow, this mechanism allows us to assign, *notionally*, a flow of 1 to (v_3, v_2) since its excess capacity is 1. Suppose this is now done. It will allow us to change our minds and take back our original assignment of $x_{23} = 1$. Defining $x_{32} = 1$, the excess capacity of (v_3, v_2) is reduced to zero. The excess capacity of (v_2, v_3) is increased from 0 to 1 and flows x_{23} and x_{32}, both 1, cancel each other out. We redefine $x_{23} = x_{32} = 0$ and we are back where we started.

This process just described allows us to modify existing flows. Let us define it formally. Assume that arc (v_i, v_j) has flow x_{ij}. If a further flow of x_{ij} is assigned to it we define

$$e_{ij} \quad \text{to become} \quad e_{ij} - x_{ij},$$

and

$$e_{ji} \quad \text{to become} \quad e_{ji} + x_{ij}.$$

Extra flow assigned to arc (v_i, v_j) with $e_{ij} > 0$ is termed *forward flow*.

Extra flow assigned in the reverse direction (because $e_{ji} > 0$) is termed *backward flow*.

It may be that there is no defined direction for the link between two nodes v_i, and v_j. In this case two oppositely directed arcs (v_i, v_j) and (v_j, v_i) are defined, each with capacity equal to the original link capacity. The notion of subtracting x_{ij} and x_{ji}, one from the other, still applies.

We now use these ideas in explaining the labelling method by applying it to the example problem. We define a label (a_i, d_i), for each node v_i, in the network other than v_1. Let

$d_i =$ the amount of extra flow which can be transported from v_1 to v_i, over and above the current flow assignment, while obeying arc capacity.

Set

$$d_i = \infty.$$

The other nodes are progressively labelled with their d_i's as the method proceeds. Set

$$x_{ij} = 0,$$

and

$$e_{ij} = u_{ij} \quad \text{for each arc } (v_i, v_j).$$

At each iteration any unlabelled point v_j, directly connected by an arc (v_i, v_j) with excess capacity $e_{ij} > 0$, to a labelled point v_i is identified.

For each identified point v_j, set

$$a_j = v_i,$$

and

$$d_j = \min\{e_{ij}, d_i\}.$$

Thus we have labelled v_j with the label (a_j, d_j), indicating that it is possible to transport an extra flow of d_j units from v_1 to v_j via v_i.

We now perform this for the example problem, illustrated in Figure 12.5. Set

$$d_1 = \infty.$$

All other nodes are unlabelled. All x_{ij} values are set to be zero, $e_{12} = 2$, and $e_{14} = 3$. We could label v_2 and v_4 with (1,2) and (1,3) respectively. Only one label is assigned at a time. The node with the smallest index is labelled first. Thus v_3 can be labelled:

$$a_3 = 2,$$

and

$$d_3 = \min\{d_2, e_{23}\} = \min\{2, 1\} = 1.$$

Therefore $(a_3, d_3) = (2, 1)$. In this way we can label v_5 as $(a_5, d_5) = (3, 1)$. Once the sink v_5, has been labelled *breakthough* has been achieved. We have now found, as displayed in the v_5 label, that we can transport a unit of flow from v_1 to v_5 via v_2 and v_3. The path of the assignment is deduced by examining the a_i's: $a_5 = 3$, so the path has a final arc of (v_3, v_5). Also $a_3 = 2$, so the penultimate arc is (v_2, v_3). The complete path is $\langle 1, 2, 3, 5 \rangle$. We now increase the flow in these arcs by $d_5 = 1$ unit:

$$x_{12} = x_{22} = x_{35} = 1.$$

The excess capacities of these arcs are reduced by the flow just assigned:

$$e_{12} = 2 - 1 = 1,$$
$$e_{23} = 1 - 1 = 0, \qquad \text{and}$$
$$e_{35} = 1 - 1 = 0.$$

The excess capacities of the oppositely-directed arcs are all increased by this flow:

$$e_{21} = 0 + 1 = 1,$$
$$e_{32} = 0 + 1 = 1, \qquad \text{and}$$
$$e_{53} = 0 + 1 = 1.$$

We have just completed one iteration of the labelling method. All labels, except for d_1, are removed and we start over again. The current flow assignment and the new excess capacities are shown in Figure 12.6.

We begin the second iteration by labelling v_2 with $(a_2, d_2) = (1, 1)$, as the excess capacity of (v_1, v_2) is now reduced to one. We cannot label v_3 as we did last time because $e_{23} = 0$. So we label v_5:

$$a_5 = 2,$$

and

$$d_5 = \min\{d_2, e_{25}\} = \min\{1, 2\} = 1.$$

Breakthrough has been achieved once more and we can augment the path $\langle a_2, a_5, 5 \rangle = \langle 1, 2, 5 \rangle$ with the flow $d_5 = 1$. Thus

$$x_{12} \quad \text{becomes} \quad x_{12} + 1 = 1 + 1 = 2$$

and

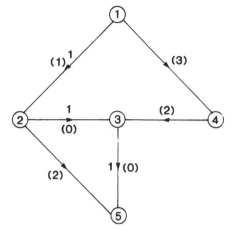

Figure 12.6 The flow after the first iteration

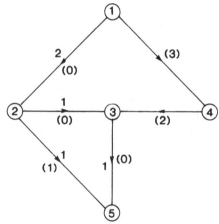

Figure 12.7 The current flow

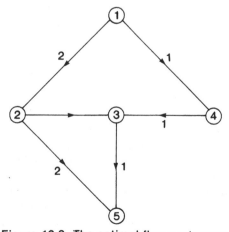

Figure 12.8 The optimal flow assignment

$$x_{25} \quad \text{becomes} \quad 1.$$

We set

$$e_{25} \quad \text{to become} \quad e_{25} - 1 = 2 - 1 = 1,$$

and

$$e_{12} \quad \text{to become} \quad e_{12} - 1 = 1 - 1 = 0.$$

The excess capacities of the oppositely-directed arcs are increased by this flow:

$$e_{21} \quad \text{becomes} \quad e_{21} + 1 = 1 + 1 = 2,$$

and

$$e_{52} \quad \text{becomes} \quad e_{52} + 1 = 0 + 1 = 1.$$

All labels are now removed. The current flow is shown in Figure 12.7.

We begin the third iteration by labelling v_4 (We cannot label v_2 as $e_{12} = 0$) with : $a_4 = 1$ and $v_4 = 3$. We can now label v_3 with $(a_3, d_3) = (4, 2)$. It now looks as if we have reached an impasse. However we have the imaginary arc (v_3, v_2) with excess capacity $e_{32} = 1 > 0$. We can use it to label v_2 with $(a_2, d_2) = (3, 1)$. This allows us to label v_3 with $(a_5, d_5) = (2, 1)$. We have found a final path $\langle 1, 4, 3, 2, 5 \rangle$, to which we can add one further unit of flow:

$$x_{14} = 1,$$
$$x_{43} = 1,$$
$$x_{32} = 1, \quad \text{and}$$
$$x_{25} \quad \text{becomes} \quad x_{25} + 1 = 1 + 1 = 2.$$

The excess capacities of the oppositely directed arcs are increased by this flow:

$$e_{41} = 0 + 1 = 1,$$
$$e_{34} = 0 + 1 = 1,$$
$$e_{23} = 0 + 1 = 1, \quad \text{and}$$
$$e_{52} = 1 + 1 = 2.$$

Now an interesting thing occurs. We have a flow of $x_{23} = 1$ in arc (v_2, v_3) and a flow of $x_{32} = 1$ in arc (v_3, v_2). These flows are now cancelled and we set $x_{23} = x_{32} = 0$. We have removed our earlier flow of $x_{23} = 1$. The final flows are shown in Figure 12.8. At the next iteration the labelling method fails to achieve breakthough. The total flow b, is 3, which is the capacity of the minimal cut $\{(v_2, v_5), (v_3, v_5)\}$. The optimal solution has been found. A formal statement of the labelling method is now given.

The Labelling Method

(1) Label the source v_1, with $(a_1, d_1) = (0, \infty)$.

(2) Set each x_{ij} value equal to that corresponding to a feasible flow. That is, one in which arc capacity and flow conservation are respected. In the event that other feasible flow assignments are available, for all arcs (v_i, v_j), both structural and introduced, set:

$x = 0$, and

$e_{ij} = u_{ij} - x_{ij}$.

(3) If there is no labelled vertex v_i, directly connected to an unlabelled point v_j by an arc (v_i, v_j), terminate the method. In this case the present flow assignment is optimal. Otherwise, continue.

(4) Find an arc (v_i, v_j), joining a labelled vertex v_i to an unlabelled vertex v_j. Set

$$a_j = v_i, \qquad \text{and}$$
$$d_j = \min\{e_{ij}, d_j\}.$$

(5) If the sink v_n, is unlabelled go back to Step (3). Otherwise go to Step (6).

(6) For each arc (v_i, v_j), on the path of labelled vertices from v_1 to v_n set

$$e_{ij} \quad \text{to become } e_{ij} - d_n, \qquad \text{and}$$
$$x_{ij} \quad \text{to become } x_{ij} + d_n.$$

For each arc (v_j, v_i), oppositely directed to arc (v_i, v_j) just identified, set

$$e_{ji} \quad \text{to become } e_{ji} + d_n.$$

If, for any pair of arcs (v_i, v_j) and (v_j, v_i), it is true that x_{ij} and x_{ji} are both positive, set

$$x_{ij} \quad \text{to become } x_{ij} - |x_{ij} - x_{ji}|,$$
$$x_{ji} \quad \text{to become } x_{ji} - |x_{ij} - x_{ji}|,$$
$$e_{ij} \quad \text{to become } e_{ij} + |x_{ij} - x_{ji}|, \qquad \text{and}$$
$$e_{ji} \quad \text{to become } e_{ji} + |x_{ij} - x_{ji}|.$$

Erase all labels except that of d_1 and return to Step (3).

Proof of Theorem 12.3

Let b equal the maximum possible source-to-sink flow found by the labelling method. The removal of the arcs in the minimal cut S, creates two separate networks. Let the sets of vertices of these two networks be V_1 and V_2. We assume that v_1, the source, is in V_1 and that v_n, the sink, is in V_2. Consider the arcs joining V_1 and V_2. The net sum of the flows between V_1 and V_2 must be b, for otherwise b would not be the maximum flow possible.

That is, for all $v_i, v_m \in V_1$ and all $v_j, v_k \in V_2$:

$$\sum x_{ij} - \sum x_{km} = b.$$

Further, for each arc (v_i, v_j), joining V_1 to V_2, we must have

$$x_{ij} = u_{ij}. \tag{12.7}$$

Otherwise v_j could be labelled with forward flow from v_i. Summing (12.7) over all arcs in C, for all $v_i \in V_1$ and all $v_j \in V_2$:

$$\sum x_{ij} = \sum u_{ij}. \tag{12.8}$$

Similarly, for each arc (v_k, v_m), joining V_2 to V_1, we must have $x_{km} = 0$. Otherwise, vertex v_k could be labelled with backwards flow from v_m. Summing over all arcs connecting V_2 to V_1,

$$\sum_{\substack{v_i \in V_1 \\ v_j \in V_2}} x_{ij} - \sum_{\substack{v_k \in V_2 \\ v_m \in V_1}} x_{km} = C(S) - 0 = b,$$

where $C(S)$ equals the capacity of S.

Thus the maximum flow equals the capacity of the minimal-cut set. As the maximum flow must be less than or equal to the capacity of any cut set and we have equality, and the result follows. ■

We now introduce a second simple transportation network problem and an algorithm for it.

The Minimum Cost Flow Problem

We define the minimum cost flow problem for a given transportation network $N = (V, A)$, with n nodes. We make the same assumptions for N as were made in Section 12.5.3. It is further assumed that there is a unit cost c_{ij}, associated with the flow in each arc $v_i v_j$, of A. It is required to assign flows to the arcs of A so that a given quantity of flow, say b, travels from the source to the sink at minimum total cost. Note that b may be strictly less than the maximum possible source-to-sink flow. Naturally, if b exceeds the capacity of the minimal cut, the problem has no feasible solution. The problem is to

$$\text{minimize } Z = \sum c_{ij} x_{ij}, \qquad \text{for all } (v_i, v_j) \in A, \tag{12.9}$$

subject to (12.2), (12.3), (12.4) and (12.5).

If $b = 1$, and

$$c_{ij} = 1, \qquad \text{for all } (v_i, v_j) \in A, \tag{12.10}$$

the problem reduces to the shortest path problem (see Section 12.5.2).

We now motivate a straightforward algorithm for the minimum cost flow problem by considering an instance of the problem based on the network in Figure 12.9. It is an extension of the labelling method, introduced in Section 12.5.3. The cost of one unit flowing along each arc is shown, along with its capacity. That is, each arc ij, in Figure 12.9 has an ordered pair (u_{ij}, c_{ij}), associated with it, where u_{ij} is its capacity and c_{ij} is its unit cost. Suppose that it is given that $b = 2$. How can the two units be transported from source v_1, to sink v_5, at minimum total cost? The conservation of flow at intermediate vertices is once again assumed.

We can build a mathematical model of the problem, which is very similar to that of the maximum-flow problem of Section 12.5.3:

$$\text{Minimize } Z = \sum_{(v_i, v_j) \in A} c_{ij} x_{ij}, \qquad \text{for all arcs}$$

subject to

$$x_{12} + x_{14} = 2,$$
$$x_{12} - x_{23} - x_{25} = 0,$$
$$x_{23} + x_{43} - x_{35} = 0,$$
$$x_{14} - x_{43} = 0,$$
$$x_{25} + x_{35} = 2,$$
$$0 \le x_{12} \le 2, \qquad 0 \le x_{23} \le 1, \qquad 0 \le x_{14} \le 3,$$
$$0 \le x_{43} \le 2, \qquad 0 \le x_{25} \le 2, \quad \text{and} \quad 0 \le x_{35} \le 1.$$

A minimum cost-flow algorithm

We now explain a simple iterative algorithm to solve the problem. More efficient algorithms are referenced in Section 12.6. At each iteration the algorithm identifies, among all the source to sink paths with excess capacity, the path with least total cost. Flow equal to the excess capacity is then added to this path. The process continues in this way until the total source-to-sink flow equals b, the given quantity to be shipped. Of course, the amount assigned in the last iteration may be strictly less than excess capacity in order to ensure that exactly b units (and not more) are shipped.

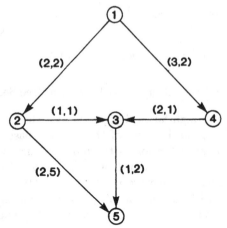

Figure 12.9 The network for the minimum-cost-flow problem

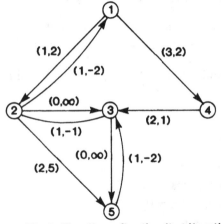

Figure 12.10 The flow after the first iteration

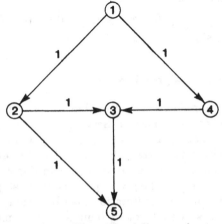

Figure 12.11 The optimal flow assignment

There is an important point which must be made about the way in which the total cost of each path is calculated. The cost of arcs oriented in the same direction as the source-to-sink path (forward flow) are added to the total cost. The cost of arcs oriented in the opposite direction to the source to sink path (backwards flow) are subtracted from the total cost. We begin by setting:

$$x_{ij} = 0, \qquad \text{for all arcs } (v_i, v_j) \in A$$

A shortest path from v_1 to v_n in Figure 12.9 is

$$\langle 1, 2, 3, 5 \rangle,$$

with a cost of:

$$c_{12} + c_{23} + c_{35} = 2 + 1 + 2 = 5.$$

This path has excess capacity because, initially,

$$e_{ij} = u_{ij} \qquad \text{for all arcs } (v_i, v_j).$$

The excess capacity of this path is one unit. This amount is assigned to the path and the same bookkeeping as for the labelling method is performed:

$$x_{12} = x_{23} = x_{35} = 1.$$

The excess capacities of these arcs are reduced by the flow just assigned:

$$e_{12} = 2 - 1 = 1,$$
$$e_{23} = 1 - 1 = 0, \qquad \text{and}$$
$$e_{35} = 1 - 1 = 0.$$

Arcs with zero excess capacity have their costs temporarily assigned to an arbitrarily large number, which we denote by ∞. The excess capacities of the oppositely directed arcs are all increased by this flow:

$$e_{21} = 0 + 1 = 1,$$
$$e_{32} = 0 + 1 = 1, \qquad \text{and}$$
$$e_{53} = 0 + 1 = 1.$$

The unit cost of the newly created arcs are defined as:

$$x_{21} = -2,$$
$$x_{32} = -1, \qquad \text{and}$$
$$x_{53} = -2,$$

that is, the negative of the unit cost of the oppositely directed arc. The current flow assignment and the new arc labels are shown in Figure 12.10.

A shortest path in Figure 12.10, from v_1 to v_n with excess capacity is

$$\langle 1, 4, 3, 2, 5 \rangle,$$

with a cost of:

$$c_{14} + c_{43} + c_{32} + c_{25} = 2 + 1 - 1 + 5 = 7.$$

This path has an excess capacity of one unit. This amount is assigned to the path. When the bookkeeping is performed, flow in arcs (2,3) and (3,2) are cancelled. We have now assigned a total of two units. As this equals b, the method is terminated. The final flows are shown in Figure 12.11.

The total cost can be calculated for paths $\langle 1, 2, 5 \rangle$ and $\langle 1, 4, 3, 5 \rangle$ as follows:

$\langle 1, 2, 5 \rangle : x_{12}c_{12} + x_{25}c_{25} = (1 \times 2) + (1 \times 5) = 7.$
$\langle 1, 4, 3, 5 \rangle : x_{14}c_{14} + x_{43}c_{43} + x_{35}c_{35} = (1 \times 2) + (1 \times 1) + (1 \times 2) = 5.$
Z (the total cost) $= 12.$

A formal statement of the method is now given.

Summary of the Minimum-Cost-Flow Algorithm

Steps

(1) Set
  $x_{ij} = 0,$
  $e_{ij} = u_{ij}$ for all arcs $(v_i, v_j),$
  $G = 0$, denoting the $v_1 - v_n$ flow assigned so far.
(2) Identify the $v_1 - v_n$ path, of least cost with positive excess capacity, E. For each arc (v_i, v_j), on this path set
  E to become $\min\{E, b - G\},$
  e_{ij} to become $e_{ij} - E,$
  x_{ij} to become $x_{ij} + E,$
  c_{ij} to become ∞ if $e_{ij} = 0$, and
  G to become $G + E.$
  For each arc (v_j, v_i), oppositely directed to arc (v_i, v_j), just identified, set
  e_{ji} to become $e_{ji} + E$ and

c_{ji} to become $-c_{ij}$.

If, for any pair of arcs (v_i, v_j) and (v_j, v_i) it happens that: $x_{ij} > 0$ and $x_{ji} > 0$, set

x_{ij} to become $x_{ij} - |x_{ij} - x_{ji}|$,

x_{ji} to become $x_{ij} - |x_{ij} - x_{ji}|$,

e_{ij} to become $e_{ij} + |x_{ij} - x_{ji}|$, and

e_{ji} to become $e_{ji} + |x_{ij} - x_{ji}|$.

(3) If $G < b$ go to Step (2). If $G = b$, terminate, as the current x_{ij} values indicate a least-cost solution.

We now discuss more advanced transportation network models and algorithms.

12.6 Transportation Networks: Advanced Models

The study of transportation network models was given an important boost when Dantzig's simplex algorithm (1963) for the linear programming problem (see Chvátal (1983)) was specialized to what has become known as the *simplex algorithm on a graph*. There are numerous presentations in the literature of successful applications of transportation network models to solve practical OR problems. These include transportation and communication networks, equipment replacement, project planning, production and inventory control, machine loading, blending, training, and financial planning. Golden and Magnanti (1977) have provided a useful bibliography of the literature on deterministic network optimization, Glover and Klingman (1977) discuss network applications in government and industry, and Assad (1978) surveys multicommodity network flow techniques. Further examples of problems formulated in terms of networks are given in the survey by Kennington and Helgason (1980). Advanced network algorithms are discussed by Tarjan (1983) and Aho et al.(1983).

12.6.1 The Capacitated Transshipment Problem (CTP)

Consider a transportation network $N = (V, A)$, in which it is desired to send a commodity between various points at minimal total cost. Each arc has a cost per unit of flow, and a capacity to accommodate flow. Each vertex v_i, is assumed to obey a flow balance law, i.e. the amount of the commodity flowing into it equals the amount flowing out of it, except for a quantity b_i. Let

$u_{ij} = $ the capacity of arc $v_i v_j$,

$x_{ij} = $ the amount of flow sent along arc $v_i v_j$,

$b_i=$ the supply of the commodity at v_i. (Each v_i for which $b < 0$ is interpreted as a demand of b_i units.)

$c_{ij}=$ the unit cost of arc v_iv_j.

The CTP problem can be formulated as,

$$\text{Minimize} \sum c_{ij}x_{ij}, \quad \text{for all } (i,j) \in A,$$

subject to

$$\sum_j x_{ij} - \sum_j x_{ji} = b_i, \quad \text{for all } v_i, v_j \in V, \tag{12.11}$$

$$0 \le x_{ij} \le u_{ij}, \quad \text{for all } (v_i, v_j) \in A. \tag{12.12}$$

A number of other models, some we have already seen, turn out to be special cases of the above model. These include: (i) the transportation problem, in which every vertex is either a source or a sink and $u_{ij} = \infty$ for all $(v_i, v_j) \in A$, (ii) the shortest path problem (Sections 12.2.4, and 12.5.2) in which $b_i = 1$ or -1 for each $v_i \in V$, and $u_{ij} = \infty$ for all $(v_i, v_j) \in A$, (iii) the maximum flow problem (Section 12.5) in which b_i is positive for exactly one $v_i \in V$, b_i is negative for exactly one other $v_i \in V$ and $b_i = 0$ otherwise, (iv) the minimum cost flow problem (Section 12.5.3) which is a combination of (i) and (iii), and finally (v) the assignment problem, (Section 12.2.5) which is a special case of (i). The CTP algorithm to be discussed here is usually so efficient that there is often no need to use a special algorithm for these specialized cases. A simple algorithm for the CTP has already been outlined in Section 12.5. The matrix form of the CTP is

$$\text{Minimize } CX \tag{12.13}$$
$$\text{subject to } AX = B, \tag{12.14}$$
$$0 \le X \le U, \tag{12.15}$$

where the matrix A is the incidence matrix for the underlying network $N = (V, A)$.

Thus the CTP is an upper-bounded linear programming problem with a very special structure. This structure gives rise to the following theorem.

Theorem 12.3 (The Basis Tree Theorem)

A set of columns of A comprise a basis if and only if the corresponding set of arcs form a spanning tree of the undirected graph derived from N.

Proof

Dantzig (1963, chapter 17), or Kennington and Helgason (1980, chapter 3).

Theorem 12.3 allows the upper bounded revised simplex algorithm (see Chvátal (1983)) to be used to solve the CTP very efficiently. This has been done by Bradley, Brown, and Graves (1977) who have devised a primal network code call GNET. The use of graph theory in modifying the simplex method is fully detailed in Kennington and Helgason (1980).

There is another method for the CTP, called the *out-of-kilter method*, which was developed by Fulkerson (1961) . Unlike the method just mentioned it was devised solely for CTP and is not a specialization of a more general method. The method begins with a feasible set of flows and a set of dual variables. It comprises two phases: one in which changes are made to flows, the other in which changes are made to the dual variables. After an iteration of each phase, an examination is made of each arc in the network. Conditions for optimality (based on the Kuhn-Tucker conditions) must be satisfied by each arc. Arcs which obey the conditions are said to be *in kilter* and those which are not, *out of kilter*. At each iteration an *out of kilter* arc has its status changed permanently to *in kilter*. When all arcs are in kilter, the optimal solution is at hand.

Computational experience comparing the two algorithms has been reported by Barr, Glover, and Klingman (1974), and by Glover and Klingman (1978). There is conflicting evidence as to which is superior.

12.6.2 The Multicommodity Network Flow Problem

This problem is similar to the last except that several different commodities are sent between the nodes of the network $N = (V, A)$. The problem can be modelled as

Minimize

$$\sum_{k=1}^{K} c^k x^k \qquad (12.16)$$

subject to

$$AX^k = B^k, k = 1, 2, \ldots, k, \qquad (12.17)$$

$$\sum_{k=1}^{K} d^k x^k \leq R, \qquad (12.18)$$

$$0 \leq x^k \leq u^k, k = 1, 2, \ldots, k, \qquad (12.19)$$

where A is once again the incidence matrix of network N. The jth component of R is the mutual arc capacity of the appropriate arc, to be shared among the k commodities. Kennington and Helgason (1980) described in detail two of the most common algorithms for the problem, which take advantage of its special structure.

12.6.3 CTP with side constraints

The following model is a specialization of the CTP in which there are further restrictions, over and above those on arc capacity and flow balance:

$$\text{Minimize } CX + EZ \tag{12.20}$$
$$\text{subject to } AX = B, \tag{12.21}$$
$$SX + PZ = R, \quad \text{and} \tag{12.22}$$
$$0 \le X \le U, \quad 0 \le Z \le V. \tag{12.23}$$

A is once again the incidence matrix of the underlying network, but S and P are arbitrary. The methods for the CTP can be modified to accommodate the side constraints (12.22). This is achieved by partitioning the basis so that a portion of it corresponds to a directed spanning tree. All calculations involving this component of the basis are executed via labelling operations rather than by matrix multiplication. (See Kennington and Helgason (1980) for full details.)

12.6.4 CTP with convex costs

The CTP model can be generalized by replacing the objective function by $g(X)$, a convex function of the arc flows X. This results in a nonlinear programming problem with special structure. It can be solved by a variety of adapted nonlinear programming techniques, each of which exploit the structure. These adoptions eliminate the need for matrix manipulation and all operations can be carried out on the network N. Kennington and Helgason (1980) present adoptions of piecewise linear approximation, and other methods.

12.6.5 The Linear Multicommodity Model

Rao and Zionts (1968) discuss the general framework in which a linear multicommodity model has been applied to a variety of transportation-allocation problems. Further applications involve the allocation of empty freight cars in railways (White and Bomberault (1969)), the routing of fuel oil tankers (Bellmore et al. (1971)), the racial desegregation of schools

(Clarke and Surkis (1968)) , the analysis of waste management systems (Panagiotakopoulos (1976)) , and problems in operation management (Elmaghraby 1970)).

12.6.6 Communications Networks

In a communications system, messages pass from one station to another via channels of given capacity. This system is often modelled by a network with flow, vertices, and arcs representing messages, stations, and channels respectively. Problems of realizability, analysis, and synthesis give rise to linear models and computer networks give rise to nonlinear models. Flows correspond to average message rates sent on different channels in store-and-forward computer networks (See Fratta et al. (1973)) . The objective is to minimize the total delay per message, given certain transmission requirements. Sometimes it is appropriate to include a concave objective function. As an example, Yaged (1973) has adopted a function to reflect the economies of scale in providing a given number of channels on a given link to satisfy flow requirements. Yaged poses a dynamic model for the problem of minimizing the net value of installation costs. Other models with a concave objective function arise in computer routing problems where capacities are treated as decision variables and costs are assigned for installing capacity. (See Fratta et al (1973).)

12.7 Summary

We have presented an introduction to graph theoretic models and solution procedures for OR models ranging from the Chinese Postman's Problem to routing in computer networks.

It must now be clear to the reader that graph theory and its offshoots, digraphs and networks, have blossomed, not only as a branch of mathematics, but also as systematic tools in OR. Graph theoretic methods afford new insights into a wide variety of OR models.

12.8 Exercises

12.1 Find the minimal spanning tree for the following edge-weight table using Kruskal's method:

	1	2	3	4	5	6	7	8	9	10
2	3									
3	4	16								
4	9	19	13							
5	8	20	1	16						
6	7	4	2	4	13					
7	6	12	5	9	3	6				
8	5	14	9	7	14	7	4			
9	4	17	14	6	5	9	5	7		
10	20	8	15	11	9	3	19	9	10	
11	15	3	20	12	19	12	6	19	15	20

12.2 If d_{12} is increased from 3 to 20 in problem 12.1, how is the optimal tree affected?

12.3 Solve problem 12.1 above using Prim's method.

12.4 Solve problem 12.2 above using Prim's method.

12.5 Devise an efficient method for solving problems of the form of Exercise 12.2.

12.6 Prove that the weight of a minimal spanning tree of any complete weighted graph is a lower bound on the length of any travelling salesman's tour of the graph.

12.7 Prove that any uniquely shortest path of a weighted graph G, must be part of any minimal spanning tree of G.

12.8 Suppose that any chord c (an edge not part of a given spanning tree T) is added to T. Prove that c has length no less than that of any edge in the cycle its addition creates.

12.9 Apply Dijkstra's method to find the shortest path from vertex 1 to vertex 10 in the graph with edge weight table given below:

	1	2	3	4	5	6	7	8	9
2	3								
3	4	6							
4	5	7	11						
5	–	9	12	9					
6	–	–	14	8	19				
7	–	–	15	12	20	27			
8	–	–	–	–	23	29	40		
9	–	–	–	–	26	30	41	50	
10	–	–	–	–	–	31	42	55	60

12.10 Find the shortest paths between all pairs of vertices 1,2,3, and 4 in the above tables using Floyd's method.

12.11 Compare the efficiency, in terms of the number of elementary arithmetical operations, in using Dijkstra's and Floyd's methods to find the shortest path between (a) a given pair of vertices in a weighted graph and (b) all pairs of vertices in a weighted graph.

12.12 Find the minimum cut in the network given in Exercise 12.9 where the entries now represent capacities. The source is vertex 1 and the sink is vertex 10.

12.13 Solve the maximum-flow problem posed in Exercise 12.12 by the labelling method.

12.14 Suppose that two nations A and B are at war. A has the option of bombing B's territory, which contains a network of roads and bridges. Nation A wishes to disrupt the flow of materials in B's territory. Using the notions of this chapter explain why it is likely to be more advantageous for A to bomb the bridges rather than the roads.

12.15 Consider Exercise 12.12 as a minimum-cost flow problem, where each arc has unit cost equal to its capacity. Assume that the maximum quantity possible is to be shipped from node 1 to node 10. Solve this problem by the method of Section 12.5.3.

12.16 Devise a method for transforming a minimum-cost flow problem with multiple sources and multiple sinks into a problem with a single source and a single sink.

12.17 Prove that the weight of a minimal spanning tree of a complete weighted graph is a lower bound on the length of a travelling salesman's tour of the graph.

13 Electrical Engineering

Across the wires the electric message came:
"He is no better, he is much the same".

Alfred Austin

Electrical energy has a wide variety of extremely important uses in today's society. Some of the design problems faced by the electrical engineer in taking advantage of these uses can be modelled in terms of digraphs. We discuss two such problems in this chapter: electrical network analysis and printed circuit board design.

13.1 Electrical Network Analysis

An electrical network is a system connecting various elements such as: resistances, inductances, voltage supplies, capacitances, and current sources. Two variables, the current variable and the voltage variable, are associated with each element. Because these variables are often functions of time, it is convenient to specify a convention of direction for them. This is best done by using a digraph (as introduced in Sections 1.2 and 7.1) to model the network. Different network elements have different physical relationships between their current and voltage variables. We shall specify some examples in this section.

269

In Section 1.1.5 we briefly touched on the two fundamental laws of electrical network theory first enunciated by G. Kirchhoff (1847). They have been universally adopted by electrical engineers in the study of networks. We shall now study electrical networks, assuming Kirchhoff's laws, with the aid of graph theory.

We begin with a formal description of electrical networks and Kirchhoff's postulates. An *electrical network* is a digraph $D = (V, A)$ which has three entities associated with each of its arcs k:

(i) An electrical element E_k,

(ii) An electrical flow (current) variable $i_k(t)$, and

(iii) A potential difference (voltage) variable $v_k(t)$.

Here t represents time. For any element E_k, these three entities are linked by a number of simple relationships.

$$(1) \quad v_k(t) = g_k(t)i_k(t) \qquad \text{(Ohm's Law for a resistor)} \qquad (13.1)$$

$$(2) \quad v_k(t) = g_k(t)\frac{di_k(t)}{dt} \qquad \text{(Inductor)} \qquad (13.2)$$

$$(3) \quad v_k(t) = g_k(t)\int_{t_0}^{t} i_k(t)dt \qquad \text{(Capacitor)} \qquad (13.3)$$

$$(4) \quad v_k(t) = f(t) \qquad \text{(Voltage supply)} \qquad (13.4)$$

Here $g_k(t)$ represents a conductance coefficient, f is a function of time, and t_0 is a starting point in time.

In general, we express the relationship between $v_k(t)$ and $i_k(t)$ as

$$v_k(t) = G_k^{-1}(i_k). \qquad (13.5)$$

These relationships imply that if the physical properties of an element E_k, are known (e.g. the function G_k^{-1}) then one of $v_k(t)$ and $i_k(t)$ can be deduced from the other.

We now state the two postulates that have come to be known as Kirchhoff's Laws.

(I) **Kirchhoff's Current Postulate (CP)**

The algebraic sum of the current variables of the arcs incident with any vertex is zero.

(II) **Kirchhoff's Voltage Postulate (VP)**

The algebraic sum of the voltage variables of the arcs of any cycle is zero.

Here the term *algebraic sum* has meaning for CP and for VP as follows:

(CP) Each current variable is added (subtracted) if its associated arc is incident away from (towards) the vertex in question.

(VP) The cycle in question is assumed to be oriented in one of the two possible directions. Each voltage variable is added (subtracted) if the direction of its associated arc is compatible (incompatible) with the cycle orientation.

The conventions of algebraic sum are illustrated in Figure 13.1, where in (a) the current variable values are shown for the arcs incident with a vertex v. According to the convention just adopted, their algebraic sum is 3 + 6 - 4 + 9 - 14 (= 0). In (b) the voltage variable values are shown for the arcs of a cycle with the orientation shown. Once again, according to our convention, their algebraic sum is 6 + 7 - 5 + 20 - 28 (= 0).

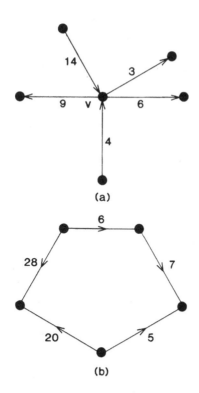

(a)

(b)

Figure 13.1   The algebraic sum of (a) current variables, (b) voltage variables

Consider now a relatively large electrical network connecting many elements. In such a system, the voltage and current variables of each element are dependent upon relationships such as (13.1) – (13.4) and Kirchhoff's Laws. The problem of electrical network analysis is to calculate the current and voltage variable values for all elements in a given network when some of the values have been specified. The solution of the set of simultaneous equations arising from the relationships can often be greatly simplified by the application of certain concepts and results from graph theory. We now explain the simplification process.

Many electrical networks have a large number of vertices and cycles. The solution of all the simultaneous equations involving the current and voltage variables that arise from the relationships just mentioned is a daunting task. It is more efficient to consider a carefully chosen subset of the equations from which the values of all the variables can be deduced.

This subset may be chosen by identifying a set of fundamental cycles and of fundamental cut sets based on a spanning tree of the network. (See Sections 3.6 and 7.4 for an explanation of these concepts.) If the network has m arcs, then these two sets will give rise to exactly m simultaneous equations to be solved.

Suppose an electrical network $D = (V, A)$, with n vertices and m arcs is given. Each arc has associated with it an element whose current and voltage variables are linked by a relationship of the form (13.5). A spanning tree T, of D is identified which, of course, will have $(n - 1)$ branches. We create a set of $(n - 1)$ fundamental cut-sets, one for each branch of T. Each fundamental cut-set comprises exactly one, say b, of the branches and any of the $(m - n + 1)$ chords of D (with respect to T) which join the two components of T created by the deletion of b. We create a set of $(m - n + 1)$ fundamental cycles, one for each chord of D with respect to T. Each fundamental cycle comprises exactly one, say c, chord and any of the $(n - 1)$ branches of T, which create a cycle when c is added to T. It is convenient from a computational point of view, if it is possible, to include all the arcs corresponding to variables with known values in either T or the chords of D with respect to T. That is, the arcs of all the known variables should, if possible, either be all in T or none in T.

The set of fundamental cut-sets can be used to create a set of equations corresponding to CP. The set of fundamental cycles can be used to create a set of equations corresponding to VP. In sum, we have a total of $2m$ equations:

    (i) m equations of the form (13.5), and

    (ii) $(n - 1)$ CP (fundamental cut-set) equations,

    (iii) $(m - n + 1)$ VP (fundamental cycle) equations.

Often the number of equations necessary to find values for all variables can be reduced considerably. For instance, as has been stated earlier, if either the value of the voltage or the current variable is known, along with the physical specification of an element, then the value of the other variables can be deduced immediately from the equation of the form (13.5). The use of reductions of this type, together with some simple matrix algebra, sometimes reduces the number of equations to be solved.

Let the subscript b, correspond to branches of the spanning tree T, selected and the subscript c correspond to the chords with respect to T.

Let

$$G = [G_b : G_c], \tag{13.6}$$

(The vector of operators expressed in general form in (13.5).)

$$
\begin{aligned}
K_f &= [I_{n-1} : K_c], \\
C_f &= [C_b : I_{n-1}], \\
I &= [I_b : I_c]^T \quad \text{and} \\
V &= [V_b : V_c]^T
\end{aligned}
$$

$$\tag{13.7}$$
$$\tag{13.8}$$

Note that: (i) the superscript T indicates the transpose of a vector (or matrix) and (ii) for matrix I_θ, if $\theta = n - 1$, I is an $(n-1) \times (n-1)$ identity matrix, but if $\theta = b$ or c then I is a vector of current variables.

Then (13.5), VP, and CP and can be written as:

$$I = G(V), \tag{13.9}$$
$$[I_{n-1} : K_c]I_i = 0, \quad \text{and} \tag{13.10}$$
$$[C_b : I_{n-1}]V = 0, \text{ respectively.} \tag{13.11}$$

Using (13.7) and (13.8), the equations (13.10) and (13.11) can be rewritten as:

$$I_b + K_c I_c = 0 \quad \text{and} \tag{13.12}$$
$$C_b V_b + V_c = 0. \tag{13.13}$$

That is,

$$I_b = -K_c I_c, \quad \text{and} \tag{13.14}$$
$$V_c = -C_b V_b. \tag{13.15}$$

By using (13.9) and (13.15) we have

$$I_c = -G_c(C_b V_b). \tag{13.16}$$

By using (13.14) and (13.16) we have

$$I_b = -K_c G_c(C_b V_b). \tag{13.17}$$

From (13.15), (13.16), and (13.17) it can be seen that the values of the voltage variables corresponding to the chords and all current variables can be calculated as combinations of the voltage variables corresponding to the branches. We now illustrate the above relationships by using them to deduce the values of all unknown variables associated with the electrical network in Figure 13.2 (a). Alternatively we can use the relationship, deduced in Section 7.4, that

$$C_b = -K_c^T \tag{13.18}$$

Thus (13.15), (13.16), and (13.17) can be rewritten as:

$$V_c = K_c^T V_b, \tag{13.19}$$
$$I_c = G_c(K_b^T V_b), \quad \text{and} \tag{13.20}$$
$$I_b = C_b^T G_c(C_b V_b). \tag{13.21}$$

It is also possible to base the computations on I_c, rather than V_b. We begin by rewriting (13.9) as

$$V = G^{-1}(I). \tag{13.22}$$

More specifically,

$$V_b = G^{-1}(I_b). \tag{13.23}$$

From (13.14) and (13.18) we have

$$I_b = C_b^T I_c. \tag{13.24}$$

By using (13.23) and (13.24),

$$V_b = G^{-1}(C_b^T I_c). \tag{13.25}$$

From (13.19) and (13.25),

$$V_c = K_c^T G^{-1}(C_b^T I_c). \tag{13.26}$$

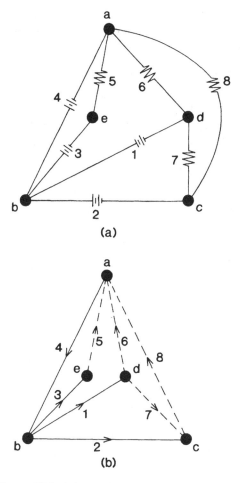

Figure 13.2 (a) an electrical network and
(b) its corresponding digraph

From (13.24) – (13.26) it can be seen that the values of the current variables corresponding to the branches and all voltage variables can be calculated as combinations of the current variables corresponding to the chords.

We now illustrate the above relationships by using them to deduce values of all unknown variables associated with the electrical network in Figure 13.2 (a). Note that 1,2,3 and 4 are voltage sources and the value of their voltage variables: v_1, v_2, v_3, and v_4 are assumed known. Components 5, 6, 7, and 8 are resistors whose physical specifications: g_5, g_6, g_7, and g_8 are assumed known.

The equations of the form of (13.9) are:

$$
\begin{bmatrix} i_5 \\ i_6 \\ i_7 \\ i_8 \end{bmatrix} = \begin{bmatrix} g_5 & 0 & 0 & 0 \\ 0 & g_6 & 0 & 0 \\ 0 & 0 & g_7 & 0 \\ 0 & 0 & 0 & g_8 \end{bmatrix} \begin{bmatrix} v_5 \\ v_6 \\ v_7 \\ v_8 \end{bmatrix}.
$$

Figure 13.2(b) displays a digraph with arcs given arbitrary orientation and labelled according to their components. A spanning tree, with solid arcs, has been selected. This tree has been used to generate the fundamental cut-set matrix K_f, and the fundamental cycle matrix C_f:

$$
\begin{array}{c}
\begin{array}{cccccccc} 1 & 2 & 3 & 4 & 5 & 6 & 7 & 8 \end{array} \\
\begin{array}{c} K_1 \\ K_2 \\ K_3 \\ K_4 \end{array}
\begin{bmatrix} 1 & 0 & 0 & 0 & 0 & -1 & -1 & 0 \\ 0 & 1 & 0 & 0 & 0 & 0 & 1 & -1 \\ 0 & 0 & 1 & 0 & -1 & 0 & 0 & 0 \\ 0 & 0 & 0 & 1 & -1 & -1 & 0 & -1 \end{bmatrix}
\begin{bmatrix} i_1 \\ i_2 \\ i_3 \\ i_4 \\ i_5 \\ i_6 \\ i_7 \\ i_8 \end{bmatrix} = 0,
\end{array} \tag{13.27}
$$

$$
\begin{array}{c}
\begin{array}{cccccccc} 1 & 2 & 3 & 4 & 5 & 6 & 7 & 8 \end{array} \\
\begin{array}{c} C_1 \\ C_2 \\ C_3 \\ C_4 \end{array}
\begin{bmatrix} 0 & 0 & 1 & 1 & 1 & 0 & 0 & 0 \\ 1 & 0 & 0 & 1 & 0 & 1 & 0 & 0 \\ 1 & -1 & 0 & 0 & 0 & 0 & 1 & 0 \\ 0 & 1 & 0 & 1 & 0 & 0 & 0 & 1 \end{bmatrix}
\begin{bmatrix} v_1 \\ v_2 \\ v_3 \\ v_4 \\ v_5 \\ v_6 \\ v_7 \\ v_8 \end{bmatrix} = 0.
\end{array} \tag{13.28}
$$

Note that the equations in (13.27) correspond to implementations of VP about the vertices: d, c, e, and a respectively. The equations in (13.28) correspond to implementations of CP about the cycles containing arcs: $\{3,4,5\}$, $\{1,2,5\}$, $\{1,2,7\}$ and $\{2,4,8\}$ respectively.

Thus

$$
K_c = \begin{bmatrix} 0 & -1 & -1 & 0 \\ 0 & 0 & 1 & -1 \\ -1 & 0 & 0 & 0 \\ -1 & -1 & 0 & -1 \end{bmatrix}, \quad \text{and} \tag{13.29}
$$

$$
C_b = \begin{bmatrix} 0 & 0 & 1 & 1 \\ 1 & 0 & 0 & 1 \\ 1 & -1 & 0 & 0 \\ 0 & 1 & 0 & 1 \end{bmatrix}. \tag{13.30}
$$

Let us assume that

$$V_b = [1, 1, 1, 1]^T. \tag{13.31}$$

Then by (13.16)

$$V_c = \begin{bmatrix} v_5 \\ v_6 \\ v_7 \\ v_8 \end{bmatrix} = - \begin{bmatrix} 0 & 0 & 1 & 1 \\ 1 & 0 & 0 & 1 \\ 1 & -1 & 0 & 0 \\ 0 & 1 & 0 & 1 \end{bmatrix} \begin{bmatrix} 1 \\ 1 \\ 1 \\ 1 \end{bmatrix} = \begin{bmatrix} -2 \\ -2 \\ 0 \\ -2 \end{bmatrix} \tag{13.32}$$

Let us assume that $(g_5, g_6, g_7, g_8) = (1, 2, 3, 4)$, that is,

$$G_c(V) = \begin{bmatrix} 1 & 0 & 0 & 0 \\ 0 & 2 & 0 & 0 \\ 0 & 0 & 3 & 0 \\ 0 & 0 & 0 & 4 \end{bmatrix} \begin{bmatrix} v_5 \\ v_6 \\ v_7 \\ v_8 \end{bmatrix} \tag{13.33}$$

Then by (13.9)

$$I_c = G_c(V_c) = \begin{bmatrix} 1 & 0 & 0 & 0 \\ 0 & 2 & 0 & 0 \\ 0 & 0 & 3 & 0 \\ 0 & 0 & 0 & 4 \end{bmatrix} \begin{bmatrix} v_5 \\ v_6 \\ v_7 \\ v_8 \end{bmatrix} \tag{13.34}$$

By (13.16) and (13.2)),

$$I_c = \begin{bmatrix} i_5 \\ i_6 \\ i_7 \\ i_8 \end{bmatrix} = - \begin{bmatrix} 1 & 0 & 0 & 0 \\ 0 & 2 & 0 & 0 \\ 0 & 0 & 3 & 0 \\ 0 & 0 & 0 & 4 \end{bmatrix} \begin{bmatrix} 0 & 0 & 1 & 1 \\ 1 & 0 & 0 & 1 \\ 1 & -1 & 0 & 0 \\ 0 & 1 & 0 & 1 \end{bmatrix} \begin{bmatrix} 1 \\ 1 \\ 1 \\ 1 \end{bmatrix} = \begin{bmatrix} -2 \\ -4 \\ 0 \\ -8 \end{bmatrix} \tag{13.35}$$

Finally, by (13.17), (13.20), (13.30) and (13.31),

$$I_b = \begin{bmatrix} i_1 \\ i_2 \\ i_3 \\ i_4 \end{bmatrix}$$

$$= - \begin{bmatrix} 0 & -1 & -1 & 0 \\ 0 & 0 & 1 & -1 \\ -1 & 0 & 0 & 0 \\ -1 & -1 & 0 & -1 \end{bmatrix} \begin{bmatrix} 1 & 0 & 0 & 0 \\ 0 & 2 & 0 & 0 \\ 0 & 0 & 3 & 0 \\ 0 & 0 & 0 & 4 \end{bmatrix} \begin{bmatrix} 0 & 0 & 1 & 1 \\ 1 & 0 & 0 & 1 \\ 1 & -1 & 0 & 0 \\ 0 & 1 & 0 & 1 \end{bmatrix} \begin{bmatrix} 1 \\ 1 \\ 1 \\ 1 \end{bmatrix}$$

$$= \begin{bmatrix} -4 \\ -8 \\ -2 \\ -14 \end{bmatrix}. \tag{13.36}$$

A negative value for the variable of any arc implies a positive value of the same magnitude for the arc orientated in the opposite direction. Suppose that it is assumed that instead of the voltage variables: v_1, v_2, v_3 and v_4 being given, the current variables: i_5, i_6, i_7, and i_8 are known, with values

$$I_c = [-2, -4, 0, -8]^T. \tag{13.37}$$

Then by (13.24),

$$I_b = \begin{bmatrix} i_1 \\ i_2 \\ i_3 \\ i_4 \end{bmatrix} = \begin{bmatrix} 0 & 1 & 1 & 0 \\ 0 & 0 & -1 & 1 \\ 1 & 0 & 0 & 0 \\ 1 & 1 & 0 & 1 \end{bmatrix} \begin{bmatrix} -2 \\ -4 \\ 0 \\ -8 \end{bmatrix} = \begin{bmatrix} -4 \\ -8 \\ -2 \\ -14 \end{bmatrix}. \tag{13.38}$$

Suppose that $(g_1, g_2, g_3, g_4) = -(4, 8, 2, 14)$, i.e.

$$G_b(V_b) = - \begin{bmatrix} 4 & 0 & 0 & 0 \\ 0 & 8 & 0 & 0 \\ 0 & 0 & 2 & 0 \\ 0 & 0 & 0 & 14 \end{bmatrix} \begin{bmatrix} v_1 \\ v_2 \\ v_3 \\ v_4 \end{bmatrix}. \tag{13.39}$$

That is,

$$G_b^{-1}(I_b) = - \begin{bmatrix} \frac{1}{4} & 0 & 0 & 0 \\ 0 & \frac{1}{8} & 0 & 0 \\ 0 & 0 & \frac{1}{2} & 0 \\ 0 & 0 & 0 & \frac{1}{4} \end{bmatrix} \begin{bmatrix} i_1 \\ i_2 \\ i_3 \\ i_4 \end{bmatrix}. \tag{13.40}$$

Then by (13.23)

$$V_b = G_b^{-1}(I_b) = - \begin{bmatrix} \frac{1}{4} & 0 & 0 & 0 \\ 0 & \frac{1}{8} & 0 & 0 \\ 0 & 0 & \frac{1}{2} & 0 \\ 0 & 0 & 0 & \frac{1}{4} \end{bmatrix} \begin{bmatrix} i_1 \\ i_2 \\ i_3 \\ i_4 \end{bmatrix}. \tag{13.41}$$

By (13.25) and (13.38)

$$V_b = \begin{bmatrix} v_1 \\ v_2 \\ v_3 \\ v_4 \end{bmatrix} = - \begin{bmatrix} \frac{1}{4} & 0 & 0 & 0 \\ 0 & \frac{1}{8} & 0 & 0 \\ 0 & 0 & \frac{1}{2} & 0 \\ 0 & 0 & 0 & \frac{1}{4} \end{bmatrix} \begin{bmatrix} 0 & 1 & 1 & 0 \\ 0 & 0 & -1 & 1 \\ 1 & 0 & 0 & 0 \\ 1 & 1 & 0 & 1 \end{bmatrix} \begin{bmatrix} -2 \\ -4 \\ 0 \\ -8 \end{bmatrix} = \begin{bmatrix} -1 \\ -1 \\ -1 \\ -1 \end{bmatrix}.$$

Finally by (13.26), (13.29), (13.30) and (13.37),

$$
V_C = \begin{bmatrix} v_5 \\ v_6 \\ v_7 \\ v_8 \end{bmatrix}
$$

$$
= - \begin{bmatrix} 0 & 0 & -1 & -1 \\ -1 & 0 & 0 & -1 \\ -1 & 1 & 0 & 0 \\ 0 & -1 & 0 & -1 \end{bmatrix} \begin{bmatrix} \frac{1}{4} & 0 & 0 & 0 \\ 0 & \frac{1}{8} & 0 & 0 \\ 0 & 0 & \frac{1}{2} & 0 \\ 0 & 0 & 0 & \frac{1}{4} \end{bmatrix} \begin{bmatrix} 0 & 1 & 1 & 0 \\ 0 & 0 & -1 & 1 \\ 1 & 0 & 0 & 0 \\ 1 & 1 & 0 & 1 \end{bmatrix} \begin{bmatrix} -2 \\ -4 \\ 0 \\ -8 \end{bmatrix}
$$

$$
= \begin{bmatrix} -2 \\ -2 \\ 0 \\ -2 \end{bmatrix}.
$$

13.2 Printed Circuit Board Design
(Based on Foulds, Perara, and Robinson (1978).)

In this age of solid state physics, the printed circuit is an invaluable device in modern electronics. A printed circuit, comprises a non-conducting board and a number of terminals. It is necessary in printed circuit board design to link various pairs of terminals by plating a conducting material on the board. Because this material is not insulated, it is vital that the conductors meet only at terminals, otherwise short-circuiting would occur. We assume that a number of terminals and the required connections between them are given. The problem is to locate the terminals and lay out conductors on the board so that the conductors do not intersect, except at their terminals.

We can model this problem in terms of graph theory as follows. The terminals are represented by the vertices of a graph and each conductor by an edge which is incident with the vertices representing the terminals that it is to connect. Then the problem can be solved (by laying out the conductors on one side of the board) if and only if the corresponding graph is planar. See Chapter 5 for a discussion of graph planarity and related concepts. An algorithm for testing whether or not a given graph is planar is given in Section 9.5.3.

If the graph is nonplanar, but has thickness two, then it is possible to design the board successfully by laying out the conductors on both sides of the board. The two planar subgraphs of the original graph are laid out, one to each side. Two terminals corresponding to each vertex are located directly opposite each other, one on each side, directly connected by a conductor.

If the graph has thickness greater than two, it is necessary to stack two more more boards in parallel planes, close together, with insulating material in between. The above process is repeated on a number of sides of boards equal to the thickness of the underlying graph.

There is also the possibility of allowing conductors to pass through holes in the board, from one side to the other, at any location which is not a terminal. If this possibility is feasible, then any terminal adjacency requirement can be accomodated on two sides of a single board. It is of interest of find such a layout which minimizes the number of holes (at locations other than terminals) that must be drilled. The number of holes will be equal to, at most, twice the crossing number of the underlying graph.

This is so because, a crude upper bound on the maximum number of holes that need to be drilled can be attained by the following strategy. Begin by locating all the terminals on one side (called the upper side) of the board only. Lay out the conductors on that side corresponding to a drawing in the plane of the underlying graph which possesses the minimum possible number of crossings. Very close to a crossing, drill two holes, one on either side of the crossing. Pass one of the conductors that crosses through one of the holes, along the other (under) side of the board, and back up through the second hole. Remove the short segment of conductor on the upper side between the two holes. Repeat this process for all crossings. This process, in effect, removes all crossings by having one of the offending conductors at each crossing rerouted under its crossing partner.

The problem of calculating the crossing number of a graph is NP-hard. As the number of vertices in a complete graph increases, the task of calculating the crossing number quickly goes outside the bounds of reasonable computing time. This means there probably does not exist an efficient algorithm that will design printed circuit boards for a large number of conductors with the minimum number of crossings.

13.2.1 A printed circuit board design heuristic

We present a heuristic solution procedure for laying out a printed circuit on two sides of a board of nonconducting material. It is assumed that the conductors are uninsulated and they must not overlap, except at an intended junction. All junction points are located on one side of the board in a given configuration. Each junction is then connected by a conductor passing directly through the board to another junction on the other side. Each of these pairs of connected junction points is considered to be one node. It is assumed that certain nodes must be connected to each other without conductors overlapping. To make this possible in complex layouts, holes are drilled in the plate. One conductor is passed through each hole.

A conductor may be passed through more than one hole. The problem is to minimize the number of holes that have to be drilled. This problem is described more fully by Busacker and Saaty (1965). The problem can be viewed in terms of graph theory. A given layout can be represented by an electrical network with nodes connected by (undirected) links. The pairs of junction points originally connected opposite one another are represented by nodes. The conductors are represented by links. The holes drilled are represented by additional nodes. When an additional node is introduced on a link, it splits the link into two new links, which are labelled A and B respectively, corresponding to the two different sides of the plate. All links not thus split are also labelled either A or B. Links on side $A(B)$ of the plate are labelled $A(B)$. The problem is to lay out the links and locate the minimum number of additional nodes so as to ensure that there is no overlap of links with the same label.

The solution procedure presented uses a computer subroutine in conjunction with an iconic model of the board comprising a pair of plastic sheets, representing the two sides of the plate, overlaid one on top of the other. Nodes and links were drawn on them with felt-tip pens. The subroutine is based on a method of Nicholson (1968), which attempts to minimize the number of overlaps which may occur when a given circuit is laid out on one side of a plate. Nicholson's method is discussed in the next section. Owens (1971), has calculated an upper bound on the number of overlaps if no holes are drilled and has developed a model yielding this bound for complete graphs.

13.2.2   A permutation procedure

Nicholson's method is used to lay out the links of a network on one side of the board so that the number of crossings is close to a minimum. The representation scheme used by Nicholson consists of laying out the nodes of the network in a straight line and drawing the connections between nodes as semicircles above and below the line.

Nicholson's method begins by constructing an initial permutation. The node with the largest number of connections is placed in position 1. The node next selected is the node which has the most connections with those already placed in the permutation. This node is placed in the position that would cause the smallest increase in the number of crossings. Its connections are drawn as semicircles above or below the node line depending on which path causes fewer crossings. When all nodes have been positioned, the initial solution has been obtained. This solution is then modified systematically to try to reduce the number of crossings. The node whose connections cause the most number of crossings is identified. This node is moved from its present position to the position which gives the highest reduction in the number of crossings. Nodes are removed until no further improvements can be made.

Nicholson's method may not necessarily produce an optimal layout of links, because this procedure does not interchange groups of two or more nodes to minimize the number of crossings. The following solved problem illustrates Nicholson's method.

The example network consists of eight nodes and the connections between pairs of nodes are given below.

1-2	2-1	3-1	4-2	5-3	6-1	7-1	8-2
1-3	2-4	3-5	4-5	5-4	6-3	7-3	8-3
1-6	2-8	3-6	4-8	5-7	6-7	7-5	8-4
1-7		3-7		5-8		7-6	8-5
		3-8					

Figure 13.3 shows how Nicholson's method builds up the layout. The positions are numbered in parentheses. Node 3 is involved with the most connections and so is inserted in position 1 in Figure 13.3(a). Node 1 is arbitrarily selected as the node with the most connections with those already placed. It is then inserted in position 1, node 3 going to position 2. The link between them is drawn as in Figure 13.3.(b). Nodes 6 and 7 are both connected to all the nodes so far placed. Node 7 is arbitrarily selected next, followed by node 6. These selections are shown in Figures 13.3(c) and 13.3(d). Nodes 5 and 8 are placed in Figures 13.3(e) and 13.3(f). Now nodes 2 and 4 are both connected to two of the nodes already placed. Node 4 is arbitrarily chosen next followed by node 2. These selections are shown in Figures 13.3(g) and 13.3(h). It can be seen in Figure 13.3(b) that there is a crossing between connections 5-7, and 3-6. This can be eliminated by moving node 7 to position 5. Thus nodes, 6, 1, and 3 are now in positions 6, 7, and 8 respectively, and nodes, 2, 4, 8, and 5 are still in positions 1, 2, 3, and 4 respectively, as shown in Figure 13.3(c). No further reductions are possible by resiting a single node and the layout given in Figure 13.3(i) is the final one produced by this procedure.

13.2.3 A solution procedure

The set of links of the network are initially divided into three subsets.

A : a set of links that are laid out on the upper side,

B : a set of links that are laid out on the lower side, and

C : a set of links that are removed from the lower side to prevent crossings.

(Later some of the links of C will be assigned to a fourth set D.)

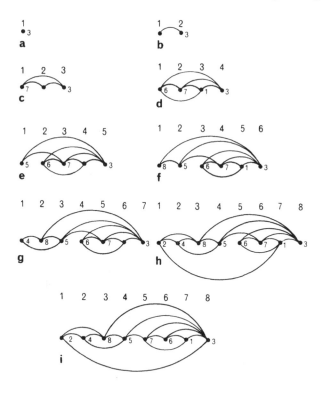

Figure 13.3 The Nicholson procedure

The purpose of this procedure is to attempt to minimize the number of holes that must be drilled in a board of nonconducting material, so that the links of the network can be laid out on the two sides of the board in such a way that no overlap occurs. There is no guarantee, however, that solutions produced by the procedure will be optimal in the sense that an absolute minimum number of holes will be required to be drilled. Initially, all the links of the network are assigned to set A. Nicholson's method is used to obtain a permutation that attempts to minimize the number of crossings between links of A. If no crossings occur, the present configuration is optimal. Otherwise the link causing the highest number of crossings is removed from A and assigned to B. The number of crossings in the remaining links is adjusted and the link causing the highest number of crossings is again removed from A and assigned to B.

This is continued until there are no crossings on the upper side. Then Nicholson's method is used to find a permutation that attempts to minimize the number of crossings when laying out the links of B. If no crossings occur, the present configuration is optimal.

Otherwise the link causing the highest number of crossings is removed from B and assigned to C. The number of crossings in the remaining links in B is adjusted and the link causing the highest number of crossings is again removed from B and assigned to C. (If no crossings occur, the present configuration is optimal.) This is continued until there are no crossings on the lower side. The set of links of the network has now been partitioned into at most three sets: A, B, and C.

An attempt is then made to reduce the cardinality of C. The link in C, say c_1, which was the last link in C to be assigned from A to B is identified. This link is removed from C and assigned to A. Nicholson's method is again used to try to minimize the number of crossings when the links of A are laid out. If no crossings occur c_1 is left in A. If crossings occur however, c_1 is removed from A and assigned to B. Again Nicholson's method is used to try to minimize the number of crossings when the links of B are laid out. If no crossings occur, c_1 is left in B. If crossings occur however, c_1 is removed from B and assigned to a new set D. The next link of C, which was the last link to be originally assigned from A to B, is removed from C and assigned to A, and the above process is repeated until C is null. The permutations found for the final assignment of links to sets A and B are then used to lay out the circuit on the plate. The links of set D will each require at least one hole and are laid out using a special subroutine.

13.2.4 Circuit layout

Every junction point of the given network has been assigned a given position on one side of the board, with a corresponding junction point directly opposite on the other side. The final permutation of nodes and links for sets A and B correspond to the upper and lower sides of the plate. The links of sets A and B are now laid out. To do this, a line which does not intersect itself is passed through the nodes on the upper side, in the order that is given in the final permutation of A. The links above and below the node line of the upper side are placed in the same way as they are placed in the final permutation of A. This is repeated on the lower side by passing a nonintersecting line through the nodes in the order given in the final permutation of B. The links above and below the node line of the lower side are placed in the same way as they are placed in the final permutation of B.

Finally the links of D must be routed so that the number of holes that must be drilled is a minimum. A special algorithm (termed algorithm L) for inserting new connections which have to be passed from one side of the plate to the other is used for this purpose. The links in D are laid out as given in step 17 of the complete procedure (see Section 13.2.5).

Step 17 is explained by referring to the example shown in Figure 13.4, in which the dashed links are those on the upper side of the plate and the solid on the lower, corresponding to sets A and B respectively. The letters in the various regions into which these lines divide the page form no part of the process as such but are used as an aid to the description.

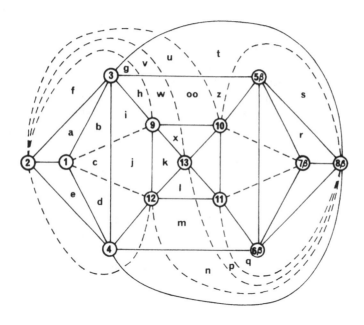

Figure 13.4 The example used to explain step 17

Suppose that D consists of the links: $\{1,5\}, \{1,6\}, \{1,7\}, \{1,8\}, \{2,5\},$ $\{2,7\}, \{3,6\}, \{3,7\}$. Step 17 begins by choosing a node α, then algorithm L is used to insert all the links from α. To choose α, we find the number of crossings that would be caused if each link is put in. Consider link $\{1,5\}$.

On the upper side it could be inserted crossing two other upper-side links, for example $\{2, 13\}$ and $\{7, 10\}$. The minimum number of crossings on the lower side is also two, namely $\{3, 4\}$, and $\{3, 9\}$. To each link is assigned the lesser of the two numbers obtained.

Each vertex is given the sum of the crossings of each edge of D associated with it. Each link has a route involving only one crossing on one side at least, except for $\{1, 5\}$ and $\{1, 6\}$, which need two on each side. The totals for the nodes are thus:

Node	1	2	3	4	5	6	7	8
Total	6	2	2	0	3	3	3	1

Hence $\alpha = 1$ is set and in the first pass through algorithm L, routes are found for links from 1 to each of 5, 6, 7, 8. These four are labelled β.

Algorithm L is now begun. The level is set to 0, colour 1 to upper and colour 2 to lower, so that until instructions are met to the contrary, colour instructions are interpreted as the side that has been set. Using subroutine 1, small zeros are put in each region: a, b, c, d, and e adjacent to 1, and are joined to 1 by arcs directed towards 1. Growing outwards from 1, upper-colouring zeros are put in each region that can be reached by crossing a lower-colour link. The regions that can be reached are completely determined, but the routes are not. For instance, the zero in g could equally be linked to the circle in f as to h. At the end of this stage, all regions within the upper-colour circle through 2, 12, 8, 13, 2 have zeros.

It is found that 8 could have been reached without any crossing, and the link is inserted along that route. As any region incident with 5, 6 or 7 has not been labelled, step 5 is next and the operation is repeated with the colours reversed. Because of triangle 2–3–4, the lower colour cannot be *grown* usefully.

The level is now raised to 1 and the colours are again interchanged. As no region has only lower-colour zeros, no progress can be made on step 11, but step 14 allows one of lower-colour to be put in every region which contains only a zero of upper-colour, and step 16 brings lower-colour ones into regions incident with 5 and 6. To reach 7 requires another change. The routes to be taken by the links follow the arrows in Figure 13.5, with a hole (filled circle) to be drilled wherever there is a change of side. The routes are:

8–1: Upperside all the way through n, m, l, k, j, c.

7–1: Upperside through r to s, then through the hole to the lower side, across t and u into v, then back to upper side, across w, x, k, j, c to 1.

6-1: Lower side across q, p, through hole in n and then on upper side parallel to 8-1.

5-1: Lower side across y, z, a, and through hole in w, and then on upper side parallel to 7-1.

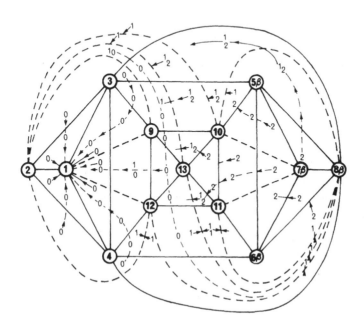

Figure 13.5 The example used to explain algorithm L

The other members of D can be inserted in a similar way.

13.2.5 The complete procedure

Step

1 Set: A = the set of all links of the given network,

$B = 0$,

$C = 0$, and

$D = 0$.

2 (a) Use the permutation procedure to find a layout of links of A on the upper side.

(b) If no crossings occur go to step 15.

3 Identify the link in A ,say a, which causes the most crossings.

4 Set A to be $A\backslash\{a\}$

Set B to be $B \cup \{a\}$.

5 Repeat steps 3 and 4 until no crossings occur on the upper side.

6 (a) Use the permutation procedure to find a layout of links on the lower side.

(b) If no crossings occur go to step 15.

7 Identify the link in B, say b, which causes the most crossings.

8 Set B to be $B\backslash\{b\}$.

Set C to be $C \cup \{b\}$.

9 Repeat steps 7 and 8 until no crossings occur on the lower side.

10 Identify the link in C, say C_1, which was the last link of C to be transferred from A to B in step 4.

11 If $C = 0$, go to step 14.

Otherwise set C to be $c\backslash\{C_1\}$.

Set A to be $A \cup \{C_1\}$.

12 Use the permutation procedure to find a layout of links of A on the upper side.

If no crossings occur go to step 10.

If crossings occur remove C_1 from A.

Let B become $B \cup \{C_1\}$.

13 Use the permutation procedure to find a layout of links of B on the lower side.

If no crossings occur go to step 10.

If crossings occur remove C_1 from B.

Let D become $D \cup \{C_1\}$.

Go to step 10.

14 Use the permutations for the final assignment of links to A and B to lay out the links of sets A and B on sides A and B respectively.

15 (a) Pass a nonintersecting line through the nodes (when they are set out in the given configuration) on the upper side of the plate in the order that is given in the final permutation of A.

(b) Place the links above and below the node line in the same way as they are placed in the final permutation of A.

16 If step 15 was reached from step 2b terminate, otherwise continue.

(a) Pass a nonintersecting line through the nodes (when they are set out in the given configuration) on the lower side of the plate in the order that is given in the final permutation of B.

(b) Place the links above and below the node line in the same way as they are placed in the final permutation of B.

(c) If Step 15 was reached from 6b, terminate. Otherwise continue.

17 The links of D are laid out as follows.

(a) Identify the node in D whose links cause the most number of crossings. Let this node be denoted by α and the nodes to which it is connected by β,

(b) Perform algorithm L,

(c) Remove links laid out by algorithm L from D,

(d) If $D = 0$ terminate, otherwise go to step 17a.

13.2.6 Conclusions

The procedure is suitable to be implemented on a *computer aided design (CAD)* system. The CAD system can be used to implement Nicholson's subroutine while the designer would physically lay out the network using a model. Depending on the situation facing the designer, there may be many physical constraints that have to be taken into account.

13.3 Summary

In this chapter we have seen how the concepts of a fundamental cycle, and a cut-set, along with their related digraph matrices, are of direct utility in electrical circuit analysis. We have also seen how the concept of planarity and related notions are useful in printed circuit board design.

13.4 Exercises

13.1 Repeat the analysis, as far as possible, for the electrical network in Figure 13.2, where the spanning tree comprises: arcs 4, 5, 6, and 7. Assume that the first three of these arcs represent 10 volt batteries and all the other arcs in the network represent 20 ohm resistors. Identify as many current and potential difference variable values in the network as possible.

13.2 Devise an electrical engineering application for each of the algorithms presented in Chapter 12.

13.3*Attempt to devise an upper bound on the crossing number for the complete bipartite graph $K_{m:n}$. (Hint: Experiment first with small values for m and n.)

13.4 If holes can be drilled anywhere in a board, what is the minimum number of holes that must be drilled if a printed circuit corresponding to K_{10} is to be laid out on one side of the board?

13.5 Devise a layout that achieves the minimum number of holes calculated in Exercise 13.4 by implementing the algorithm of Section 13.2.

14 Industrial Engineering

If you have great talents, industry will improve them.

Sir Joshua Reynolds

In this chapter we discuss applications of graph theory in industrial engineering. The applications are concerned with production planning and control, and with the design of the layout of the physical facilities of some system.

14.1 Production planning and control

One of the most common uses of digraphs in industrial engineering concerns project planning. The digraph depicting the precedence relationships is, strictly speaking, a network but it is not clear how the theory of network flows aids the analysis. There are a number of advocates for each of the two approaches: activity-oriented digraphs with vertices representing activities, and event-oriented digraphs with vertices representing events. The latter representation has the advantage that activity numbers indicate sequential relationships which makes it the more efficient for solution by computer. However this often means that dummy activities have to be introduced. The use of digraphs in project planning, involving what are called CPM and PERT networks, is discussed in Robinson and Foulds (1980).

291

Apart from the actual scheduling of activities it is often useful to construct the digraph (which is the precedence diagram) for the purpose of assembly line balancing. One common problem in this area requires the tasks to be arranged in a series of work stations so as to minimize the number of stations required without violating the precedence constraints. Chachra et al. (1979) show how the theory of digraphs can be applied to analyse this problem. They also go on to consider various other problems in industrial engineering, such as determining how many subcomponents must be fabricated to make a given number of final products, the determination of which process is best to make a given product, goods shipment scheduling (the transportation problem), and various organizational problems. They represent each of these problems in terms of digraphs and each model is a useful way to visualize the problem. However it is not clear how the theory of digraphs can be used to solve the problems.

14.2 Facilities Layout

14.2.1 Introduction

The industrial management problem of facilities planning contains the important subproblem of facilities layout. It is concerned with designing the layout of a system of physical facilities, such as buildings on a plane site, or machines on a manufacturing shop floor. The ultimate aim of the facilities layout project is to produce a plan depicting the relative positions of the facilities which are to be laid out in order to optimize some measure of the performance of the system. Typical measures are: the total travel distance of workers and goods, the sum of the desirability ratings of adjacent pairs of facilities, and the total production cost. Although the traditional scenario of facilities layout is the industrial plant, the existing layout models and techniques have application in the design of other institutions such as: police stations, civic complexes, airports, hospitals, and so on.

14.2.2 A Graph Theoretic Layout Model

It is assumed that there exists a *relationship chart* in the form of a table whose entries represent the desirability of locating a relevant pair of facilities adjacently. These entries are usually letters representing some objective measure of the desirability of the associated adjacency. We assume that they can be, and have been, converted into real numbers, termed *closeness ratings*. The higher the desirability, the larger the closeness rating. Let

$r_{uv} = $ the closeness rating for siting facilities u and v adjacently,

$n = $ the number of facilities,

$V = $ the set of facilities, and

$E =$ the set of all adjacencies, i.e. the set of all unordered pairs of facilities which can be sited next to each other.

It may be that some of the elements in E represent infeasible adjacencies and some represent compulsory adjacencies. The relevant r_{uv} values should be set to appropriate levels in these cases.

In order to facilitate understanding, we now consider a specific layout example. Consider the office block floor plan shown in Figure 14.1. Each room or facility has been assigned the vertex of a graph. The region exterior to the building has also been assigned a vertex. We treat the exterior region as just another facility because it may be convenient to rate the desirability of locating each facility adjacent to the exterior wall. We define two facilities to be adjacent if and only if their boundaries are coincident for at least a given positive length. If two facilities: f_1 and f_2, are adjacent their associated vertices v_1 and v_2, are joined by an edge $v_1 v_2$, whose geometric equivalent is a line which intersects only the common boundary of f_1 and f_2. The resulting graph depicts the adjacency structure of the layout. It is of necessity a planar graph. (See Chapter 5 for an introduction to planarity.) We can now pose a graph theoretic formulation of the problem. Let (V, E) be a weighted graph with weight function d, defined by $d(uv) = r_{uv}$, $uv \in E$.

Then the problem is to

Model 1

$$\text{Maximize} \quad \sum_{uv \in E} r_{uv} x_{uv} \qquad (14.1)$$

subject to:

$$(V, E') \quad \text{is a planar graph}, \qquad (14.2)$$
$$x_{uv} = 0 \text{ or } 1, \quad \text{for all} \quad u, v \in V, \qquad (14.3)$$
$$E' = \{uv : x_{uv} = 1, uv \in E\}, \qquad (14.4)$$

where

$$x_{uv} = 1, \text{ if facilities } u \text{ and } v \text{ are located adjacently},$$
$$0, \text{ otherwise.}$$

It can be seen that the objective is to maximize the sum of the ratings of adjacently located pairs of facilities. It is assumed that this maximizes the desirability rating of the layout, which may be proportional to the number of trips between facilities saved by adjacent location.

In practice it is usual to eliminate from E all the edges which have negative weights as there always exists an optimal solution which does not contain

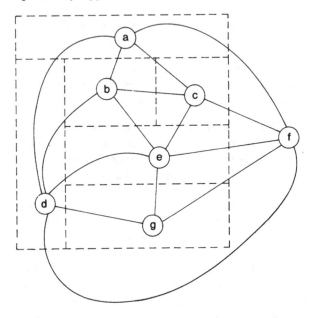

Figure 14.1 A block plan and its adjacency graph

any of them. The problem is one of finding the maximum weight planar subgraph of a given weighted graph. (The complexity of this problem is discussed in Section 9.3.) Usually at least one optimal solution will be a maximally planar subgraph because usually the edge weights are assumed to be all nonnegative. (This assumes that sufficiently many closeness ratings are nonnegative which will usually be the case in practical situations.) We turn now to solution procedures for the model.

14.2.3 An Algorithm for Model 1

The procedure to be introduced is a branch and bound enumeration approach. The general principles of branch and bound enumeration are first introduced. Some of the first introductions to the branch and bound approach to appear in the open literature were given by Balas (1965) and Lawler and Wood (1966). Later applications, in a graph theoretic context have been discussed by numerous authors including Christofides (1975), Minieka (1978), and Winston (1991). The branch and bound method is a sequential technique for solving integer programming problems which produces a decision tree. The first iteration produces the node at which the tree is rooted. Subsequent iterations produce a number of new nodes which are connected to the existing tree by arcs which all emanate from one existing node. A set of decisions representing a partial solution which, for the present problem, comprise some of the edges of a planar subgraph is associated with each node along with a bound on the value of any feasible solution which contains this set. A solution routine calculates a bound on the values of all feasible solutions to the problem and associates it with the first node. A partitioning routine creates a number of partial decisions, represented in the tree by a number of nodes which all stem from the first node. The solution routine calculates one bound for each new node in the tree. An elimination routine prunes a node from the tree if it can be shown that any feasible solution associated with it cannot be part of any optimal solution. The method continues generating new nodes at each iteration. Termination occurs when finally the optimal solution to the original problem, or evidence that no such solution exists, has been obtained. This latter evidence is provided by the elimination routine if all nodes are pruned from the tree.

Any maximally planar graph can be used to calculate an upper bound on the value of any solution in a set of feasible solutions generated by a BFS or DFS process. (These processes are discussed in Section 9.5.1.) When the partitioning process is applied to a node N, in the BFS or DFS decision tree, an edge, uv say, of the graph is identified (by a method described later) and it is either accepted as part of the solution or rejected. This creates two nodes emanating from N, denoted by uv (representing the set of feasible solutions which contain uv) and uv' (representing the set of feasible solutions which do not contain uv). The upper bound for each node is equal to the sum of the weights of the $3n - 6$ edges which can possibly belong to any solution in the set. The node selection strategy is one of selecting the node with the largest bound. The edge with the largest finite weight which can possibly belong to a solution in node N is selected for partitioning N. A path of nodes in the decision tree is fathomed when it:

(i) cannot contain any feasible solutions (detected by a negative result from a planarity testing algorithm), or

(ii) has a bound less than the value of a known feasible solution, or

(iii) the feasible solution with highest value associated with the path has a value better than that of any currently known feasible solution.

The fathoming procedure mentioned above relies on having a procedure to test whether a promising subgraph is planar or not. Unfortunately this testing is a difficult process. Hopcroft and Tarjan (1974) have presented a linear-time algorithm for graph planarity testing which is described in Section 9.5.3. However, it is difficult to program and has high overheads in terms of computing time. As there is no polynomial limit on the number of times the test may be required, it is desirable to have a series of filters that lessen the number of tests. The first two filters are trivial consequences of Kuratowski's characterization of planar graphs (see Theorem 5.4). A graph is planar if it does not contain:

(i) nine edges,

(ii) six vertices of degree at least three, or

(iii) five vertices of degree at least four.

The remaining filters involve transforming efficiently the graph to be tested (by reducing the numbers of its edges and vertices) into a smaller graph which has the same planarity status. If neither (i) or (ii) can be applied to the final graph produced, at least the Hopcroft and Tarjan test is being applied to a smaller graph. The details of the transformation are given in Foulds and Robinson (1976).

Unfortunately the method cannot solve realistically-sized problems in reasonable computing time. This comes as no surprise as the problem is NP-hard (see Section 9.3). This result raises the possibility of further analyzing the problem using polyhedral combinatorics or of devising heuristic procedures (introduced in Section 12.3) for it. As the former approach is beyond the scope of this book, we now examine graph theoretic heuristics for it.

14.2.4 Heuristic methods for model 1.

Krejcirik (1969) appears to be among the first to have recognized that graph theory could provide insights into facilities layout. In 1969 he developed the RUGR method — the first graph theoretic heuristic. Since then Moore (1976) has been involved in work which culminated in another graph theoretic heuristic applicable to the model described earlier. He had devised four strategies for the construction of the planar subgraph (V, E') using string processing grammars to ease the graph manipulation. (V, E') is then redrawn as a relationship diagram. Another graph, the dual of (V, E'), is then constructed, which is used to form a layout block plan incorporating the given areas and shapes of the facilities.

The Deltahedron heuristic

Another heuristic, due to Foulds and Robinson (1978), is based on the concept of a combinatorial deltahedron. It turns out to be useful to list, not only the vertices and edges of the subgraph being constructed, but also the regions defined by a drawing of it — here called *triangles*. This term is used because by the nature of the construction all of the regions will be bounded by three edges. The final subgraph produced can be thought of as a *polyhedron*, all of whose faces are triangles. Such a polyhedron has been called a *deltahedron* by many geometers. (See for example Cundy and Rollett (1952).) A deltahedron can be thought of as a maximal planar graph, a plane embedding of which is called a *triangulation* (see Section 5.3).

We now describe the heuristic due to Foulds and Robinson (1978) which is based on the idea of a deltahedron. Start by choosing four vertices representing four facilities. Join each pair by an edge to create K_4, the complete graph on four vertices, which in our terminology, is the *tetrahedron*. The process of selecting the tetrahedron will be described shortly. If its vertices are: a, b, c, and d; its edges are: ab, ac, ad, bc, bd, and cd; and its triangles are: abc, abd, acd, and bcd. These elements initially define the three sets: V, E', and T, respectively, of the deltahedron (V, E', T). These sets will be progressively updated until eventually they specify the final solution — subgraph (V, E'). The remaining vertices are added one at a time, choosing at each iteration a triangle in which to insert a new vertex. Suppose it is desired to insert vertex u, into triangle abc. The process is formally defined as:

$$V \quad \text{becomes} \quad V \cup \{u\},$$
$$E' \quad \text{becomes} \quad E' \cup \{au, bu, cu\}, \quad \text{and}$$
$$T \quad \text{becomes} \quad (T \setminus \{abc\}) \cup \{abu, acu, bcu\}.$$

The process is illustrated in Figure 14.2. The vertices which comprise the initial tetrahedron and the insertion selection process are as follows. Calculate for each vertex u:

$$S(u) = \sum_{v \in V} r_{uv}.$$

Arrange the vertices in order of nonincreasing S value. Ties are settled arbitrarily. The first four vertices are chosen to make up the initial tetrahedron. Vertices are then inserted, one at a time, in the order just described. Each vertex is inserted in the existing triangle which causes the largest increase in the weight (as a sum of its edge weights) of the deltahedron. That is, if u is the next vertex to be inserted, all triangles in T are examined and the one, say abc, yielding the largest sum,

$$r_{au} + r_{bu} + r_{uv},$$

is selected.

It may be possible to improve the final deltahedron with either of the following strategies. This leads to a heuristic termed *Improved Deltahedron*.

(1) Edge Replacement

In order to illustrate the process, suppose triangles ade and def appear in the final deltahedron, as in Figure 14.3. If de is deleted and af is inserted and ade and def are replaced by adf and aef, the total weight is increased by $r_{af} - r_{de}$. There is one variant necessary in this: if it is desired to delete

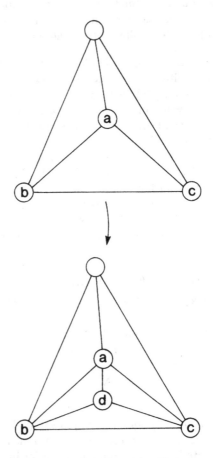

Figure 14.2 The vertex insertion process

be in Figure 14.3, the other diagonal of quadrilateral $bfeg$ is fg. But fg is already present in the deltahedron. So fg must be redrawn as the diagonal in $bfeg$, replacing be and ci is added. Thus the final changes are:

> Replace be by ci
> Replace bef, beg, cfg, and fgi by bfg, efg, cfi and cgi.

The total weight of the deltahedron is increased by $r_{ci} - r_{be}$. Such an alteration is worthwhile if this weight increase is positive. This complication raises the fear that, in terms of Figure 14.3, the new edge ci, might itself already be an edge of the deltahedron, requiring its transfer and the in-

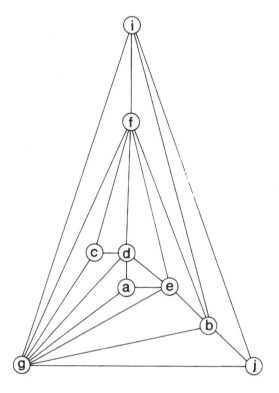

Figure 14.3 An example of the improvement
operations

sertion of another edge elsewhere, and so on, for a chain of replacements
possibly ending in the reinstatement of *be*. It has been shown by Foulds
and Robinson (1979) that this cannot happen.

(2) Vertex Relocation

A vertex of degree 3 can be removed by reversing the insertion process and
then inserting the vertex elsewhere. For example in Figure 14.3, vertex *c*
can be removed from triangle *dfg* and inserted in *bij*. It necessitates the:

> Deletion of : *cd*, *cf*, *cg*,
> Insertion of : *cb*, *ci*, *cj*,
> Deletion of : *cdf*, *cdg*, *cfg*, *bij*,   and
> Insertion of : *bci*, *bcj*, *cij*, and *dfg*.

The increase in total weight is $r_{bc} + r_{ci} + r_{cj} - r_{cd} - r_{cf} - r_{cg}$. Such an alteration is worthwhile if this weight increase is positive.

The complexity of the heuristics without the improvement phase is calculated by noting that, at the ith triangle insertion process, there are $2(i+1)$ triangles to consider. So the complexity is $O(n^2)$. When this is followed by the improvement phase, the complexity for the complete method becomes $O(n^3)$.

We now illustrate the method by using it to solve the following numerical example. This example will also be used for each of the methods to be discussed later. Computational experience with test problems for the methods is reported in Foulds, Gibbons, and Giffin (1985), Foulds and Giffin (1985) and Giffin and Foulds (1987).

Consider the following closeness rating matrix for an eight-facility layout problem.

Example 14.1

0	56	129	158	29	47	79	108
56	0	142	86	26	126	143	54
129	142	0	17	157	76	37	85
158	86	17	0	72	52	108	152
29	26	157	72	0	82	78	88
47	126	76	52	82	0	121	93
79	143	37	108	78	121	0	72
108	54	85	152	88	93	72	0

Summing the columns and reordering yields an initial deltahedron D_0, on vertices 8, 4, 3, 7, (with weight 471) and an insertion order of: 2, 1, 6, 5. The scores associated with inserting vertex 2 in the triangles of D_0 are:

$$D_0(8, 4, 3) : w_{28} + w_{24} + w_{23} = 282,$$
$$D_0(8, 4, 7) : 283,$$
$$D_0(8, 3, 7) : 339, \qquad \text{and}$$
$$D_0(4, 3, 7) : 371.$$

Hence we place vertex 2 in $D_0(4, 3, 7)$. The resulting triangulation D_1, now comprises triangles $\{(8,4,3), (8,4,7), (8,3,7), (2,4,3), (2,3,7), (2,4,7)\}$, and we evaluate the benefits of inserting the next vertex, which is vertex 1, in each of these. $D_1(8, 4, 3)$ is chosen, giving a score increase of 395. Similarly, vertex 6 is inserted in $D_2(2, 3, 7)$ and vertex 5 in $D_3(8, 3, 7)$ and the construction is complete. Figure 14.4 displays the final configuration D_4, whose solution score is 1883.

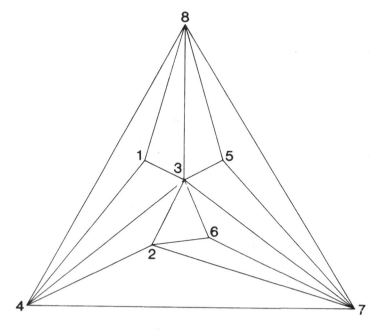

Figure 14.4

We now test whether D_4 admits improvements. The general procedure is to first test the benefit of relocating each vertex of degree three, and then to test each edge not currently in the solution as a candidate for edge replacement, choosing the better alternative at each stage. For the sake of efficiency, it proves expedient to test only $O(n)$ of the complementary edges at each pass.

In the present example no improvement is possible through vertex relocation, but the following sequence of edge replacement proves worthwhile:

edge deleted	edge added	improvement
37	56	45
34	12	39
57	68	15
		−
		99
		−

No further improvements are now possible to the resulting deltahedron whose solution score is 1982. In fact, this represents the *optimal* solution, as can be shown using the method discussed in Section 14.2.3.

A Case Study (This section is based on the M.Sc. thesis of H.V. Tran, University of Canterbury, New Zealand.)

A university has recently decided to build a new law library. The floor is to be a 12,000 square feet rectangle with sides approximately 140 and 86 feet. The facilities of the library are:

Facility	Area (Square ft.)
Catalogue	120
Issue Desk	240
Reference Desk	100
Library Staff Work Area	1,300
Copying Room	300
New Publications Display	350
Private Graduate Study Cells (6)	180
Discussion Rooms (2)	500
Foyer	340
	3,430

It is agreed to design the library to hold a projected 41,000 publications which are estimated to require an area of 4,100 sq. ft. This area is unusually large because law publications are abnormally bulky. This is anticipated to be a steady state area where weeding of obsolete publications and relocation in a warehouse of seldom-used volumes match the acquisition of new material. It will be some time before the library acquires 41,000 publications and then has to remove unpopular volumes to make way for new purchases. Until that happens, it is planned to use the extra space for additional seats and desks.

This leaves enough room for 150 seats, yielding a 1:4 ratio of numbers of seats to the total number of students. The 150 seats and the shelves of publications were divided up into blocks, as shown in Figure 14.5. The publications were divided into four categories: reports and statutes (2085 sq. ft), journals and periodicals (1226 sq. ft.), textbooks (736 sq. ft), and encyclopedias and miscellaneous digests (123 sq. ft).

The problem is one of producing a scale diagram of the layout where each facility has the correct area. The layout should be efficient in the sense

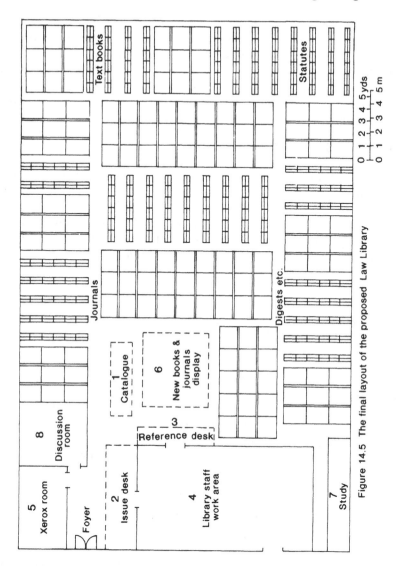

Figure 14.5 The final layout of the proposed Law Library

that it attempts to minimize travel between the various facilities and is convenient for all users, including library staff. The approach attempts to maximize travel saved by the adjacent location of facilities. So the model of Section 14.2.2 can be used and solved by the deltahedron method of the previous section.

Estimations of the traffic (in terms of the number of trips by people) between each pair of facilities were made by means of a questionnaire, distributed to librarians, law faculty, and students. These people were asked to estimate the number of trips they make between each pair of facilities in a typical week. As a result of this the r_{uv} values for Model 1 were derived. The deltahedron heuristic, when applied to this data produced a triangulation. The final layout is shown in Figure 14.5.

The Wheel Expansion Heuristic

The next heuristic is due to Eades et al. (1982). In order to explain it, we need some further graph theoretic notions. Let $G = (V, E)$ be a graph with $uv \in E$. We can define a new graph, denoted by G/uv, obtained by making u and v coincident in a drawing of G while retaining the incidence structure of G. Loops are then removed and multiple edges are replaced by a single edge. G/uv is called the *uv contraction of G*.

Theorem 14.1

If $G = (V, E)$ is a 3-connected graph and $|V| > 4$, then there exists $e \in E$ such that G/e is 3-connected.

Proof

The reader is referred to the proof by Tomassen (1980).

As maximal planar graphs are 3-connected, and the uv contraction process clearly preserves the maximal planarity property, the following result, similar to Theorem 14.1 can be proved for maximally planar graphs.

Corollary 14.1

If $G = (V, G)$ is maximally planar and $|V| > 4$, then there exists $e \in E$ such that G/e is maximally planar.

We now define an operation on maximally planar graphs that is somewhat the reverse of the uv-contraction process. Identify a vertex, say v, of degree at least four (one will exist if the graph has more vertices than the tetrahedron). Remove v and its incident edges. Replace it by two adjacent vertices say, v_1 and v_2, in the region, say R, created by the removal of v. Triangulate this region by joining v_1 and v_2 to the vertices of the boundary of R, ensuring that both v_1 and v_2 are joined to at least two of these vertices. This operation, termed *vertex expansion*, is illustrated in Figure 14.6. A *wheel* is a graph in which one vertex, called the *hub*, is joined to each of the other vertices of W by an edge, all these other vertices forming a cycle, called the *rim*. When the edge expansion process is applied to the hub of a wheel in a maximally planar graph it is termed *wheel expansion*, as it creates an extra wheel in the graph. This is because Corollary 14.1 can be used to show that:

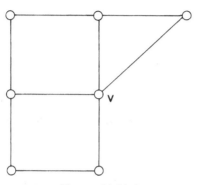

Figure 14.6(a)

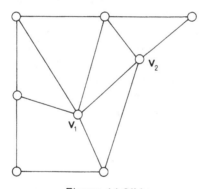

Figure 14.6(b)

Corollary 14.2

Every maximally planar graph can be obtained from K_4, by a sequence of wheel expansions.

The heuristic expands the initial K_4 by the wheel expansion process until a subgraph spanning V is obtained. The vertices of the K_4 are chosen as follows. First choose the edge in G, say uv, of largest weight. Then choose vertex w so that triangle uvw has the highest possible weight. The vertex x is chosen so as to maximize the weight of the K_4 induced by: u, v, w, and x. The general step of the heuristic is to add a vertex to the maximally planar subgraph constructed so far by wheel expansion, producing a new maximally planar graph. This is always possible because of the following theorem.

Theorem 14.2

A graph $G = (V, E)$ is maximally planar $\iff$

(i) $m = 3n - 6$ and

(ii) Each vertex $x \in V$ is the hub of a unique wheel W_x, which has exactly the set of vertices adjacent to it as its rim.

Proof

The reader is referred to the proof by Skupien (1966).

Let the vertex to be added be y and the hub to be split be x. Two vertices h and k on the rim of W_x are identified. W_x is expanded about x to create a new wheel W_y, such that h and k lie on the rim of both W_x and W_y. This is shown in Figure 14.7. Note also, that x, y, h, and k are chosen so as to make the new subgraph as heavy as possible. Note that the vertices x and y must have degree at least three but may have far higher degrees. Eades et al. (1982) show that the wheel expansion heuristic has a complexity of $O(n^4)$. However, because the degrees of the vertices are bounded, the running time of the heuristic on randomly generated problems in practice is of the order of n^3. We now illustrate the heuristic by using it to solve Example 14.1.

The greedy choice of the initial K_4 is (1, 4, 7, 8) with weight 677 (compare this to the score of 471 achieved using the deltahedron heuristic). From now on we expand the graph so that it is as heavy as possible at each iteration. The first expansion is made with vertex 4 as the hub, vertex 2 as y, $h = 7$ and $k = 1$. Now vertices 1 and 7 are adjacent on the rim of W_4. Here the wheel-expansion operation is equivalent to a deltahedron vertex insertion, namely inserting vertex 2 in triangle (4,1,7), for an increase in weight 285. This is similarly true for the next two expansions:

$$x = 4, \quad y = 6, \quad h = 7, \quad k = 2, \quad weight = 299.$$
$$x = 4, \quad y = 3, \quad h = 2, \quad k = 1, \quad weight = 288.$$

At this stage our construction may be depicted as in Figure 14.8 (a).

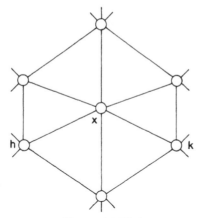

Figure 14.7(a)

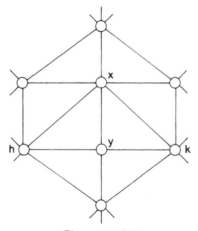

Figure 14.7(b)

The final expansion, with

$$x = 4, \quad y = 5, \quad h = 7, \quad k = 1,$$

does not have h and k adjacent on W_x, and hence the general expansion procedure is required. This involves

(i) deleting edges (x, p), where p lies on W_x between h and k,

(ii) adding edges (y, p) (p as in (i)), (y, h), (y, k), and

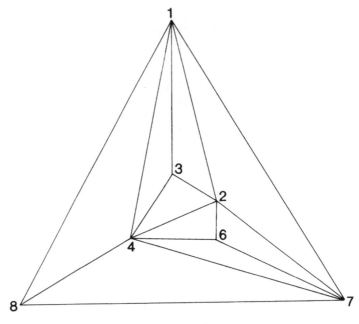

Figure 14.8(a)

(iii) adding edge (x, y).

For the example, these steps are:

   (i) delete (4,6), (4,2), (4,3),

  (ii) add (5,6), (5,2), (5,3), (5,7), (5,1), and

 (iii) add (4,5).

These modifications yield the deltahedron of Figure 14.8(b), which has an associated weight of 1838.

The Greedy Heuristic

Foulds and Giffin (1985) present another heuristic which is conceptually simpler than the one just described. It starts by ordering the edges in decreasing order of weight. Each edge is examined in this order and is accepted as part of the subgraph being constructed unless its addition creates a nonplanar subgraph.

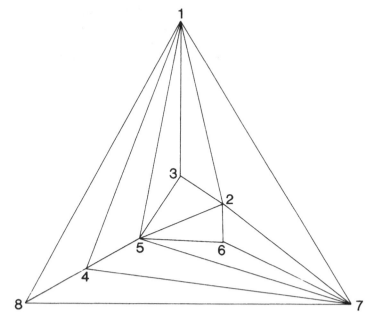

Figure 14.8(b)

In this latter case it is rejected and the next edge on the list is examined. Once a maximally planar subgraph spanning V has been built up the heuristic is terminated. The same strategies for testing for planarity, as described in Section 14.2.3, are used here. The complexity of this heuristic is $O(n^3)$.

Applying the greedy heuristic to Example 14.1, we initially make use of the filters described in Section 14.2.3. This implies that the first eight most highly weighted edges may be added, that is:

$$\{14, 35, 48, 27, 23, 13, 26, 67\}.$$

Associated with this graph are the vertex degrees: 2,3,3,2,1,2,2,1, so that the next two candidate edges, $\{1, 8\}$ and $\{4, 7\}$ may also be added without the need for planarity testing. However, edge 68 now induces a degree set of: 3,3,3,3,1,3,3,3, with 7 vertices of degree at least 3, implying the subsequent need for the Hopcroft and Tarjan test. No edges are in fact rejected until the 16th, and 17th. The graph of Figure 14.9 represents the construction thus far, showing that the only feasible candidates are edges

$\{3,6\}$, $\{7,8\}$, and $\{1,2\}$, the 18th, 20th and 21st edges in the ordered list, respectively. The weight of the completed triangulation is 1982. The greedy approach has thus achieved the optimal solution. We note that the improvement strategies described in Section 14.2.4.2 could equally be applied to the greedy heuristic.

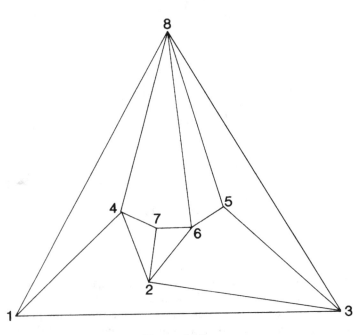

Figure 14.9

14.2.5 Refined Graph Theoretic Models

The N-Boundary Model (Based on Giffin and Foulds (1987).)

The graph theoretic model described in Section 14.2.4 assumes that no credit is to be given for the location of a pair of facilities that are somewhat near each other but not actually adjacent. It would be useful to devise a model which relaxes this assumption. This is now done.

Consider the office block layout in Figure 14.1. For each pair of facilities we can calculate the least number of boundaries that must be crossed in order to travel from one to the other. If there are n facilities Grunbaum

(1967) showed that at most $[(n - 2)/3] + 1$ boundaries must be crossed in travelling between any pair. (Here $[n]$ is the integral part of n.)

It is assumed that there is a series of relationship tables, $R_1, R_2, \ldots, R_M$, available which summarise the desirability of location of facilities relative to each other, where:

$$M = [\frac{n - 2}{3}] + 1, \qquad \text{and}$$
$$R_k = [r_{uv}^k], \qquad k = 1, 2, \ldots, M,$$

where

r_{uv}^k = the benefit (expressed as a real number) of locating facilities u and v such that k boundaries must be crossed in order to travel between them.

Usually (but not necessarily) for a given uv pair, $r_{uv}^k > r_{uv}^h$ for $k < h$. We can use the values r^k to calculate a score for any given layout.

The model assumes that the building is to be laid out by specifying the number of boundaries between each pair of facilities. The objective is to find a layout which maximizes the sum of benefits taken over all pairs of facilities, the benefit for pair u,v being defined as r_{uv}^k if they are separated by exactly k boundaries.

In terms of the graph defining the adjacency structure of the layout (as illustrated in Figure 14.1), facilities u and v, are k boundaries apart if the shortest path (in terms of the least number of edges) between vertices u and v is k.

In graph-theoretic terms, the objective is to find a triangulation T, on the n vertices representing the n facilities, which maximizes the sum of pairwise benefits. Formally, we wish to:

Model 2

$$\text{Maximize} \quad \sum_{\substack{u,v \in V \\ u \neq v}} \sum_{k=1}^{M} r_{uv} x_{uv}^k \tag{14.5},$$

subject to:

$$T = (V, E) \quad \text{is a triangulation,} \tag{14.6}$$

$$\sum_{k=1}^{M} x_{uv}^k = 1 \quad \text{where } u, v \in V, \quad \text{for } u \neq v,$$

where:

$x_{uv}^k = 1$, if the shortest path between vertices u and v has k edges,

0, otherwise,

and

$$E = \{uv : x_{uv}^1 = 1, u, v \in V\}.$$

Note that **Model 1** is a special case of **Model 2** with $M = 1$. The values x_{uv}^1 which equal one correspond to the edges of T and define which pairs of facilities are actually adjacent in the layout.

There is another graph theoretic way of viewing the problem. Let a table R_k, define the weights of a complete weighted graph G_k, on n vertices representing the n facilities, for $k = 1, 2, \ldots, M$. We wish to create a new complete weighted graph G', of maximum total weight, made up of edges (and corresponding weights) from the G_k's. The edges in G' which come from G_1 form a triangulation T, and edge uv, in G_k, $k > 1$ is in G' if and only if the shortest path from u to v in T has k edges. Once T has been found it can be used as a prescription for which pairs of facilities are to be adjacent in the layout.

Naturally any modified version of the heuristic must have a mechanism for calculating shortest paths (in the sense of the least number of edges) in the deltahedra progressively constructed. It is efficient to have an updating, rather than a start-over, subroutine in recalculating shortest paths when a new vertex is inserted in a triangle. That is, it is desirable to make use of the relevant shortest path matrix before the insertion. A shortest path algorithm due to Dantzig (1967) will be used for our purposes and is described next. Other shortest path algorithms have been discussed in Section 12.5.2.

Let $C_1, \ldots, C_n$ be a sequence of matrices, where C_i is an $i \times i$ matrix containing the length of the shortest paths for vertices: $1, 2, \ldots, i$. That is,

$$C_i = (c_i(k, j))_{i \times i},$$

where $c_i(k, j) = $ the length of the shortest path between vertices k and j.

Let

$$c_{i+1}(k, j) = \min_{\substack{k \\ 1 \leq k \leq i}} \{c_i(j, k) + c_i(k, i + 1)\}, \qquad j = 1, 2, \ldots, i,$$

and

$$c_{i+1}(i + 1, j) = \min_{\substack{k \\ 1 \leq k \leq i}} \{c_i(i + 1, k) + c_i(k, j)\}, \qquad j = 1, 2, \ldots, i.$$

Then

$$c_{i+1}(j,k) = \min\{c_i(j,k), c_{i+1}(j,i+1) + c_{i+1}(i+1,k)\},$$
$$j = 1, 2, \ldots, i, k = 1, 2, \ldots, i.$$

The above relationships define an iterative approach to calculating C_{i+1} induced by the addition of the $(i+1)$th vertex to the triangulation. They can be used to update the shortest paths efficiently, along the lines suggested by Cheston (1976) and used by Cheston and Corneil (1982). Suppose that we have C_i for an arbitrary graph $G_i = (V_i, E_i)$. Suppose now that, without loss of generality, vertex $i+1$ ($\notin V_i$) is added to G_i and is adjacent to vertices: x, y, and z, say.

Then

> **for each** $v \in V$ **do**
> $$c_{i+1}(v, i+1) := \min_{\substack{j \\ j=x,y,z}} \{c_i(v,j) + 1\}$$

(because $c_{i+1}(j, i+1) = 1$, $j = x, y, z$.)

and

> **for each** $u \in V_i$
> **for each** $v \in V \backslash \{u\}$ **do**
> $$c_i(u, v) = \min\{c_i(u, v), c_{i+1}(u, i+1) + c_{i+1}(v, i+1)\}.$$

This procedure has complexity $O(n^2)$. However, as we are dealing solely with triangulations, we can make some simplifications. In this case c_{i+1} can be found as:

(i) $c_{i+1}(i+1, x) = c_{i+1}(i+1, y) = c_{i+1}(i+1, z) = 1$,
$c_{i+1}(x, i+1) = c_{i+1}(y, i+1) = c_{i+1}(z, i+1) = 1$.

(ii) $c_{i+1}(i+1, j) = \min\{c_i(x, j), c_i(y, j), c_i(z, j)\} + 1$, $j \neq x, z, i+1$,
$c_{i+1}(j, i+1) = c_{i+1}(i+1, j)$.

(This is because the shortest path from j to $i+1$ must pass through exactly one of x, y, or z.)

(iii) $c_{i+1}(j, k) = c_i(j, k)$ $j, k \neq i+1$

(This is because the existing shortest paths are not altered by the addition of vertex $i+1$).

C_{i+1} is therefore obtained by adding one row and one column to C_i, as defined in (i), (ii), and (iii).

The modified deltahedron heuristic is called the *N-boundary deltahedron*. It is the same as the deltahedron heuristic except that the objective function of Model 1 is replaced by the objective function of Model 2 in order to calculate the triangle into which to insert each new vertex. Each evaluation of the objective function is facilitated by (i), (ii) and (iii). The N-boundary deltahedron heuristic has complexity $O(n^3)$. We now illustrate the method by applying it to Example 14.1. In Example 14.1 there are 8 facilities, so M (the maximum number of boundaries) is 3. Hence we need to generate relationship tables R_1, R_2 and R_3. Arbitrarily, we set $[R_2] = 0.55[R_1]$, $[R_3] = 0.1[R_1]$, where R_1 is the table of Example 14.1.

The initial tetrahedron and order of insertion are D_0, the same as that for the deltahedron method. The initial shortest path matrix C_4, is

$$
\begin{array}{c}
 \\
8 \\
4 \\
3 \\
7
\end{array}
\begin{array}{cccc}
8 & 4 & 3 & 7 \\
\left[\begin{array}{cccc}
0 & 1 & 1 & 1 \\
1 & 0 & 1 & 1 \\
1 & 1 & 0 & 1 \\
1 & 1 & 1 & 0
\end{array}\right].
\end{array}
$$

(For completeness we assume the existence of a homomorphism ϕ such that $\phi(i) = k$ if vertex i is inserted at the kth iteration).

First we consider placing vertex 2 in triangle $D_0(8, 4, 3)$.

The induced shortest paths are then

$$
c_5(2, 8) = c_5(2, 4) = c_5(2, 3) = 1, \qquad \text{and}
$$
$$
c_5(2, 7) = \min(1, 1, 1) + 1 = 2,
$$

and the corresponding benefit is

$$
r_{28}^1 + r_{24}^1 + r_{23}^1 + r_{27}^2 = 54 + 86 + 142 + 77 = 359.
$$

For the other triangles, the benefits are:

$$
\begin{aligned}
&D_0(8, 4, 7) : 358, \\
&D_0(8, 3, 7) : 386, \qquad \text{and} \\
&D_0(4, 3, 7) : 401.
\end{aligned}
$$

Hence, vertex 2 is inserted into $D_0(4, 3, 7)$ and the new shortest path matrix C_5 is

$$
\begin{array}{c}
8 \\
4 \\
3 \\
7 \\
2
\end{array}
\begin{array}{ccccc}
8 & 4 & 3 & 7 & 2 \\
\left[\begin{array}{ccccc}
0 & 1 & 1 & 1 & 2 \\
1 & 0 & 1 & 1 & 1 \\
1 & 1 & 0 & 1 & 1 \\
1 & 1 & 1 & 0 & 1 \\
2 & 1 & 1 & 1 & 0
\end{array}\right].
\end{array}
$$

Continuing, we insert as follows:

vertex 1 in $D_1(8,4,3)$,
vertex 6 in $D_1(2,3,7)$,
vertex 5 in $D_1(8,3,7)$,

giving an N-boundary solution benefit of 2208 and a triangulation T, of weight 1883. T is in fact the triangulation of Figure 14.4, and displays a phenomenon frequently observed when using this unconstrained formulation (see Giffin and Foulds (1987)). The degree of (at least) one vertex is $n-1$. Although not a problem in such a small example, for larger values of n (≥ 15), the existence of a vertex of degree $(n-1)$ in the adjacency graph can lead to difficulties in finding a feasible layout, so its occurrence should be minimized. This can be achieved by simply constraining the maximum degree of any vertex, and allowing this limit to be exceeded only if construction cannot otherwise continue.

For the above example, setting the degree constraint at 6 (the lowest value at which a solution is attainable) reduces the solution benefit to 2169.

Minimizing Total Transportation Cost (Based on Foulds and Giffin (1985).)

The previous graph theoretic models assume that trips are saved by the adjacent location of facilities and partially saved by relatively near location. The models have the flaw that there is no recognition of the physical distance between facilities in the actual layout. It would seem useful to incorporate into these models the notion of the minimization of total transportation cost. To this end we define for a given layout L:

$w_{uv} = $ the product of the cost per unit distance travelled and the number of trips made per time period between facilities u and v,

$d_{uv}^L = $ the distance travelled between facilities u and v in layout L,

$a_u = $ the area of facility u,

$A = (a_u)_{l \times n}$, and

$L(A) = $ the set of feasible layouts where each has area a_u for facility u.

Then the problem is to

Model 3

$$\underset{L \in L(A)}{\text{Minimize}} \quad T(L) = \sum_{u=1}^{n} \sum_{r=u+1}^{n} w_{uv} d_{uv}^L. \qquad (14.7)$$

The ideas of the preceding sections have been used to develop a solution method for Model 3, as explained in the next section.

The Super Deltahedron Method

The method produces an adjacency graph G, (a specification of which pairs of facilities are adjacent) which can be transformed into a layout L, with a relatively low $T(L)$ value. The transformation from G to L is discussed in Section 14.2.6. Basically Super Deltahedron is identical to N-Boundary Deltahedron except for the following alterations:

(i) When deciding which new vertex and which existing triangle to choose at the kth insertion phase, the combination which minimizes

$$\sum_{\substack{v \\ v \notin V_k}} \sum_{\substack{u \\ u \notin V_k}} w_{uv} \hat{d}_{uv}^L \tag{14.8}$$

is selected. Here V_k is the set of vertices of the triangulation $T_k = (V_k, E_k)$, constructed at the kth iteration. The rationale for this is that existing shortest paths between elements in V_k will not change and need not affect the choice of insertion.

(ii) In order to evaluate (14.8), estimates of each d_{uv}^L value, denoted by $\hat{d}_{uv}^L$, must be made. This is done as follows. The shortest path (in terms of the least number of edges) between u and v in T_k is found, say $\langle u, b_1, b_2, \ldots, b_M, v \rangle$, where $b_1, b_2, \ldots, b_M \in V_k$. Then d_{uv}^L is approximated by

$$\hat{d}_{uv}^L = \tfrac{1}{2}(a_u)^{\frac{1}{2}} + \sum_{i=1}^{M}(a_{b_i})^{\frac{1}{2}} + \tfrac{1}{2}(a_v)^{\frac{1}{2}}. \tag{14.9}$$

The rationale for this is that in the absence of additional information, it is assumed that each facility is to have a square shape in the final layout. The length of each side of facility i is $a_i^{\frac{1}{2}}$. Assuming rectilinear travel between the centroids of facilities u and v, the *relative* total distance travelled can be successfully approximated as in (14.9). The point is that although it is not necessarily true that $d_{uv}^L = \hat{d}_{uv}^L$ for $u - v$ pairs, (14.9) is useful in finding relative distances in order to make the best choice for insertion. The above version of the Super Deltahedron heuristic has complexity $O(n^4)$.

Rather than using the greedy approach to vertex/triangle selection, we may also fix the insertion order in the same way as for the deltahedron heuristic. This reduces the computational complexity to $O(n^3)$, and gives a procedure termed *fixed order*.

We now illustrate both versions by using them to solve Example 14.1. Let $[w_{uv}]$ be the matrix of Example 14.1, and $A = \{36, 144, 196, 64, 36, 64, 100, 144\}$. The corresponding facility side-lengths are then $(6, 12, 14, 8, 8, 10, 12.)$ In order to produce a practical adjacency graph, we again choose to limit the maximum degree of any vertex i, at a level proportional to $a_i/\sum a_i$. This constraint set is $\{6,9,10,7,6,7,8,9\}$, which is consistent with the levels used earlier, but somewhat superfluous for our 8-facility problem.

The greedy tetrahedron choice again yields: 1, 4, 8, and 7 with a corresponding solution value of

$$7w_{14} + 9w_{18} + 8w_{17} + 10w_{48} + 9w_{47} + 11w_{87} = 5994,$$

and a rectilinear shortest path matrix of:

$$
\begin{array}{c|cccc}
 & 1 & 4 & 8 & 7 \\
\hline
1 & 0 & 7 & 9 & 8 \\
4 & 7 & 0 & 10 & 9 \\
8 & 9 & 10 & 0 & 11 \\
7 & 8 & 9 & 11 & 0
\end{array}.
$$

We now consider inserting each of the remaining vertices: 2, 3, 5, and 6 in each of the triangles: (1,4,7), (1,4,8), (1,7,8), and (4,7,8). For example, 5 in (4, 7, 8):

$$d_{54} = \frac{1}{2}a_5^{\frac{1}{2}} + a_4^{\frac{1}{2}} = \frac{1}{2}(6+8) = 7,$$

$$\hat{d}_{57} = \frac{1}{2}a_5^{\frac{1}{2}}a_7^{\frac{1}{2}} = 8,$$

$$\hat{d}_{58} = \frac{1}{2}a_5^{\frac{1}{2}}a_8^{\frac{1}{2}} = 9,$$

$$\hat{d}_{51} = \min(\hat{d}_{54} + \hat{d}_{41}, \hat{d}_{57} + \hat{d}_{71}, \hat{d}_{58} + \hat{d}_{51})$$
$$= \min(7+7, 8+8, 9+9)$$
$$= 14.$$

This yields a solution score of

$$7w_{54} + 8w_{57} + 9w_{58} + 14w_{51} = 2248.$$

In fact, this is the best choice available. The shortest path matrix is then updated to

$$
\begin{array}{c|ccccc}
 & 1 & 4 & 8 & 7 & 5 \\
\hline
1 & 0 & 7 & 9 & 8 & 14 \\
4 & 7 & 0 & 10 & 9 & 7 \\
8 & 9 & 10 & 0 & 11 & 9 \\
7 & 8 & 9 & 11 & 0 & 8 \\
5 & 14 & 7 & 9 & 8 & 0
\end{array}.
$$

and the triangle set is modified. Continuing, we

> insert vertex 6 in triangle (5,8,7),
> insert vertex 2 in triangle (6,5,7), and
> insert vertex 3 in triangle (2,6,7),

for an overall solution score of 29593, and a vertex degree set of $\{3, 4, 3, 4, 5, 5, 7, 5\}$.

Under the fixed-order rationale, the initial tetrahedron is: 8, 4, 3, and 7 and the insertion order is 2, 1, 6, 5. The steps followed are then:

vertex	insertion triangle	score
2	(4,3,7)	5359
1	(8,4,3)	5584
6	(2,4,7)	6588
5	(8,3,7)	6620

leading to a solution score of 29171, a slight improvement on the greedy value. The corresponding vertex degree set is $\{3, 4, 6, 6, 3, 3, 6, 5\}$.

14.2.6 Drawing a block plan to scale

A compiled Pascal microcomputer program called LayoutManager (Foulds (1993)) converts the adjacency graph of a proposed layout into a scale block plan with given areas for the facilities.

14.3 Summary

In this chapter we have discussed applications of graph theory to the industrial engineering problems of production planning and control, and to facilities layout. No algorithms exist for guaranteeing optimality of the facilities layout problem in reasonable time for a realistic number of facilities, so it is necessary to develop efficient procedures that produce solutions approximating the optimum. We have described five graph theoretic heuristic methods for three different formulations of the problem.

Where the objective is to maximize the sum of the relationship ratings between adjacently located pairs of facilities, the *Deltahedron* and *Greedy* methods proved the most successful, uniformly outperforming *Wheel Expansion* for problems with fewer than 50 facilities. For larger examples however, *Deltahedron* requires significantly less processing time to produce a solution of similar quality to that of *Greedy*. But, given the insignificance of computing expense in comparison with the cost of the actual layout, it is recommended that both methods be used on any problem, and the better solution chosen.

Model 1 does not take account of the benefit gained from having two facilities nearby, but not actually adjacent. Providing this extension leads to a second model where we calculate the shortest path between any pair of facilities in the adjacency graph in terms of the least number of boundaries that must be crossed. Benefits gained in location are then proportional to this distance. We call this heuristic method *N-Boundary-Deltahedron*, as it uses the *Deltahedron* construction mechanisms. Performance of this technique is encouraging, both in terms of efficiency and solution quality, but care must be taken to avoid constructing an adjacency graph with an impractical corresponding layout.

None of the models mentioned above incorporate the physical sizes of the facilities; our third formulation does so under the objective of minimizing the total transportation cost. We assume that each facility may be represented by a square of given dimension, and that inter-facility movement is rectilinear. Then the construction mechanism is analogous to that of *N-Boundary-Deltahedron*. This gives rise to the heuristic method *Super Deltahedron*. Both examined variations of the basic method usually produce practical layout guidelines.

Once the adjacency graph has been constructed via any of the techniques, it remains to convert it into the corresponding block layout design that best reflects the preferred relationships. Here the most important breakthrough is the development of a further graph theoretic algorithm which will turn an adjacency graph into a scale block plan. This means that the layout problem can be solved by a two-phase approach: adjacency, then design. The splitting of the problem into two straightforward subproblems has meant that the graph theoretic approach is now a powerful addition to the industrial engineer's tool kit.

14.4 Exercises

14.1 Consider a facilities layout problem with seven facilities: 1,2,3,4,5,6, and 7. The desirability of locating the various pairs of facilities is given in the following table.

Rating	Pairs
Necessary	(1,3)
Very important	(3,5), (4,7), (6,7)
Important	(1,5), (2,5), (3,4)
Neutral	(1,2), (1,4), (1,6), (1,7), (2,3), (2,6), (2,7), (3,6), (3,7)
Unimportant	(4,5),(4,6), (5,6), (5,7)
Forbidden	(2,4)

The facilities are all to have equal area and the rating of each facility with exterior region is neutral. Prepare an adjacency graph and, by constructing its geometric dual, draw a block plan which takes into account the given ratings.

14.2 Consider the following relationship chart for a six-facility layout problem, in terms of tonnes per day of material to be moved between the various facilities.

	B	C	D	E	F
A	3	7	6	0	9
B		8	8	1	8
C			3	2	5
D				6	0
E					5

The facilities are all to have equal area. Prepare a scale layout plan which minimizes the daily work (measured as tonnes × distance travelled).

14.3 Suppose that in Exercise 14.2, the flow between facilities B and C is now 3 tonnes/day. Find the new optimal layout by modifying the layout constructed in Exercise 14.2.

14.4 Solve Exercise 14.2 with the objective of maximizing the tonnes per day transported between adjacent facilities.

14.5 Suppose that in Exercise 14.4, the flow between facilities B and C is now 3 tonnes/day. Find the new optimal layout by modifying the layout constructed in Exercise 14.4.

14.6 Consider the following relationship chart for a six-facility layout problem, viewed as input data for Model 1.

Facility Areas		2	3	4	5	6
500 m²	1	2	0	0	0	0
400 m²	2		1	0	0	2
200 m²	3			2	4	0
250 m²	4				3	4
450 m²	5					4
100 m²	6					

Draw a layout, with facilities drawn to scale.

14.7 Repeat Exercise 14.6 with the data viewed as input to Model 3.

14.8 Repeat Exercises 14.7 and 14.6, where R_1 is defined as the table given in each exercise. Subsequent tables: $R_2, R_3, \ldots$, are defined as

$$R_{i+1} = \frac{i}{i+1} R_i, \qquad i = 1, 2, 3, \ldots,.$$

14.9*Show that, in terms of the edge replacement process described in Section 14.2.2, that a removed edge will not be reinstated. Hint: See Foulds and Robinson (1979).

14.10*Show, in terms of Model 2, that at most $[(n-2)/3] + 1$ boundaries must be crossed. Hint: See Grunbaum (1967).

15 Science

Out of olde bokes, in good feith, cometh all this newe science.

The Parlement of Foules, Geoffrey Chaucer

In this chapter we discuss the application of graph theory in the natural sciences. This complements our discussion of the use of graphs in the social sciences, in Section 11.1. We begin with brief descriptions of problems, one in physics (the dimer problem) and some in chemistry, along with graph theoretic techniques for their analysis. The major application discussed in this chapter is in the area of biology, and concerns the construction of evolutionary trees.

15.1 Physics

In Section 1.1.5 we briefly outlined a graph theoretic application involving physical systems. This use of spanning trees and related concepts was elaborated upon in Section 13.1. There are a number of other applications of graph theory in physics, including continuum statistical mechanics (Temperley (1979)), percolation processes (Fisher and Essam (1961)), and enumeration (Heap (1966)). We choose a specialized topic from one of these applications, namely the *dimer problem* from statistical mechanics.

15.1.1 The dimer problem

It is common in physics to represent crystals by lattices in three dimensions in which crystal atoms and bonds are represented by lattice vertices and edges respectively. In studying the surface properties of a given crystal, it is often convenient to look upon the corresponding face of the crystal as a graph. One such property concerns the absorption of diatomic molecules, called *dimers*. To analyse this property, it is necessary to count the number of ways in which nearest neighbour molecules can be connected on a doubly periodic lattice (a rectangular lattice whose pairs of opposite sides are identified). Every molecule must be matched via a bond with exactly one of its neighboring molecules. This problem is equivalent to finding a dimer covering (perfect matching), in the graph representing the crystal face. Perfect matchings were discussed in Section 8.3. The problem is also equivalent to that of finding the number of ways of covering a rectangular board of squares with dominoes, each of the size of two squares.

Returning to the matching problem, a necessary and sufficient condition for the underlying graph to have a dimer covering is given in Theorem 8.9. Two dimer coverings are shown for the graph in Figure 15.1(a). For clarity, only the edges in the coverings are shown. Two graphs which do not have dimer coverings are shown in Figure 15.1(b).

Methods for the enumeration of the dimer coverings of a given graph have been reported by Fisher and Temperley (1961) and Kasteleyn (1961). These methods use the *Pfaffian*, a combinatorial sum of square, antisymmetric matrices, similar to the *permanent* and the *determinant*. The methods can be used to show that there are 36 dimer coverings of the 4×4 rectangular lattice and 12 988 816 dimer coverings of the 8×8 rectangular lattice.

15.2 Chemistry

In Section 1.4 we briefly mentioned the work of A. Cayley in enumerating certain chemical isomers by tree representation. This was not the first instance of overlap between graph theory and chemistry. Indeed, ever since Higgins (1789) drew chemical graphs, graph theoretic ideas have been studied by chemists.

These include Couper (1858) who was concerned with vertex degree, Laurent (1864) who proved Theorem 1.1, Crum Brown (1884) who represented various molecules as graphs, and Henze and Blair (1931) who used enumerative techniques to count several hydrocarbon molecules. More recently, graph theorists have analysed various problems in chemistry, including Cayley mentioned earlier. Others include : Sylvester (1909) who saw the

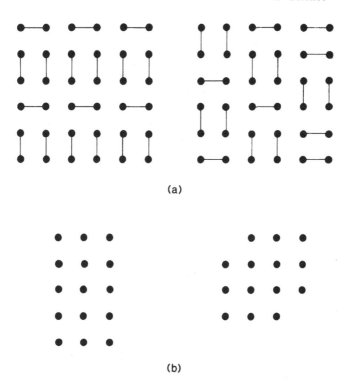

(a)

(b)

Figure 15.1. (a) Two dimer coverings and (b) graphs
which do not possess dimer coverings

correspondence between graphs and chemical formulae, Polya (1937, 1940)
who developed enumerative techniques and applied them in chemistry, and
Read and Harary (1970) who enumerated various aromatic hydrocarbons.

15.2.1 Molecule representation

In Section 1.1.4 we showed the graph theoretic representation of some
molecules with their atoms and bonds being represented by vertices and
edges respectively. For instance, the carbon, oxygen, nitrogen, and hydrogen atoms have a valence of four, three, two, and one, respectively. Thus
the corresponding vertices in the associated graphs have similar degrees.

Because multiple bonds occur between atoms, some of the representational structures are multigraphs, but not strictly graphs. To illustrate this, we consider the paraffins, with formula C_kH_{2k+2}. These molecules have $3k+2$ atoms (vertices) of which k are carbon and $(2k+2)$ are hydrogen. They have $3k+1$ bonds (edges). For example, one representative of each of the types of molecules are shown as graphs in Figure 15.2. However, the molecule *aminoacetone*, C_3H_7NO, has a multiple bond shown as a pair of parallel edges in its representation in Figure 15.3(a).

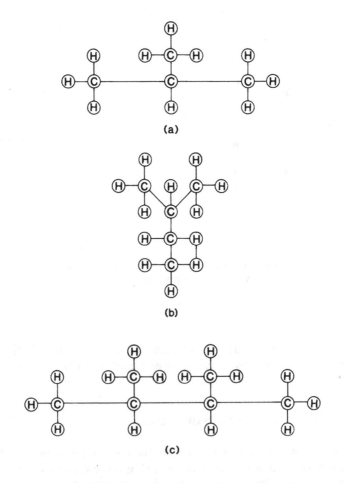

Figure 15.2. The graph theoretic representation of some paraffins: (a) butane (b) pentane and (c) hexane

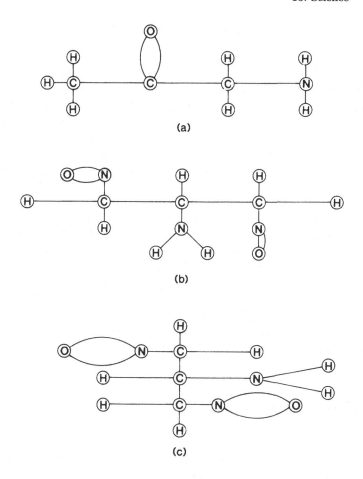

Figure 15.3   (a) C_3H_7 NO (b) and (c), two isomorphic multigraphs
representing the same compound

15.2.2 The enumeration of isomers

Such drawings of graphs are useful in that they can be used to distin-
guish the different *isomers* of a molecule. Each isomer of a molecule can
be represented by a different vertex-labelled graph with the correct vertex

degrees. For instance, Cayley (1875) showed, among other results, that there are 802 different isomers of the paraffin $C_{13}H_{28}$, each represented by a unique (non-isomorphic) tree with 41 vertices (13 with degree four and 28 with degree one). Cayley's work was extended by Henze and Blair (1931) who developed iterative techniques involving recursion formulae for enumerating the first 20 to 30 isomers of various molecules. They analysed the symmetry properties of molecules with group-theoretic techniques. This lead to a counting polynomial for enumerating the isomers. The approach has been illustrated in the enumeration of certain classes of evolutionary trees in Section 3.5, and thus will not be repeated here.

Another problem faced by the chemist is that of deciding whether two compounds, with the same chemical formula, are identical or not. Due to the graph theoretic representation approach, outlined in Section 15.2.1, this is equivalent to determining whether or not two vertex-labelled graphs are isomorphic. A discussion of the difficulty of this problem is given in Section 9.5.5. An algorithm for the problem, specialized to the graphs of chemical compounds, has been reported by Sussenguth (1965). As an illustration, it can be applied to establish that the graphs in Figure 15.3(b) and (c) represent the same compound. Further applications of graph theory in chemistry are discussed by Rouvray and Balaban (1979).

15.3 Biology

We mention one problem in genetics, before presenting, in depth, the use of graphs in evolutionary tree identification. In genetics it is often desirable to be able to test various models for the possible linkage of feasible configurations of the chemical constituents of various genes. Mutant forms of genes arise from standard forms when the connections in the gene structure are altered. It is often of interest to discover if the altered parts of two mutant genes have overlapping gene structure. A graph is a useful model in this endeavour, with vertices representing mutant genes and edges connecting them whenever the corresponding genes have overlapping modified parts. The problem of the existence of overlap can best be analysed by representing the vertices of the graph on a line with appropriately overlapping intervals. A pair of intervals of the line overlap if and only if their corresponding vertices are adjacent in the graph. Such graphs are termed *interval graphs*. Thus testing the original genetic structure is equivalent to testing its corresponding graph to see if it is an interval graph. The following theorem provides a useful characterization of interval graphs.

Theorem 15.1

A graph G is an interval graph if and only if every quadrilateral (cycle of four edges) of G has a diagonal and every odd cycle $\langle v_1, v_2, \ldots, v_k \rangle$, in the complement of G has at least one of the edges: $\{v_{k-1}, v_1\}$, $\{v_k, v_2\}$, $\{v_i, v_{i+2}\}$, for $i = 1, 2, \ldots, (k-2)$.

Proof

The proof adds little to our development and is omitted. It can be found in Gilmore and Hoffman (1964).

15.3.1 The construction of evolutionary trees   (Based, in part, on Penny et al. (1982).)

According to the theory of evolution, existing biological species have been linked in the past by common ancestors. Following the school of Charles Darwin, many researchers have displayed conjectured relationships between ancestors by drawing trees. These trees are called *phylogenies*. Predecessors of certain classes of species (vertebrates for example) have bequeathed us with a useful fossil record of themselves which can be compared with certain characteristics of similar species living today. Such analysis has produced a fair amount of agreement among scientists on the nature of the phylogenies of these classes. However, for most classes the fossil record is not sufficiently rich to be analysed in this way. Consequently, there is far less agreement on the phylogenies of these classes.

Efforts have been made to reach agreement on the phylogenetic history of these awkward classes by using an approach that is completely different from numerical taxonomy. The newer approach constructs tentative phylogenies from protein sequence data gathered by modern biochemical techniques. It appears that each generation of a species passes on to the next, a detailed account of itself, written in the chemical code of DNA, from which genes are constructed. Therefore a comparative analysis of the equivalent genes, or the proteins for which they code, from different species may reveal evidence of ancestral relationships.

With this in mind, we examine the sequence of chemical residues in five different proteins from 11 animal species. The data from each of the five proteins can then be arranged so as to fall into the simplest possible patterns of relationships between the species. This arrangement is achieved by building a graph theoretic model of evolution and solving it by combinatorial optimization techniques. In theory, 11 species could be related in over 34 million ways, but for each independent protein, the simplest pattern of relationship is remarkably similar.

Since the different proteins are independent of each other, the fact that all five give similar patterns, makes it likely that the ancestral relationship of species suggested by the patterns coincides with the reality of evolution. If the theory of evolution was not true there would be no reason for independent measures to fall into the same pattern. Furthermore, the type of pattern derived in this study fits well with those derived from such things as bone structure and the fossil record.

It comes as no surprise to most biologists that the results of this study support the theory of evolution, but it is important that the theory can now be rigorously tested. It is interesting that graph theory played a key role in making the test systematic, achievable, and unbiased.

15.3.2 Definition of the Problem

We begin this section by describing the problem of constructing a phylogeny from protein sequence data. Suppose that we wish to identify a phylogeny for a given set of species with little fossil evidence. We must use only existing species. We use information about the versions of a protein called *cytochrome c*. It seems that every living species possesses a version of this protein as it appears to be essential for respiration. Each version can be represented by a string of 312 letters chosen from the following four symbols representing what are called *nucleotides: A, C, G,* and *U*.

The problem is to construct a phylogeny which can be thought of as a weighted arborescence. Each of the given species is represented by a vertex in the phylogeny. There may be other vertices which do not have such labels. These represent either other existing species which were not included in the given set, or extinct species. The weight of each arc in the phylogeny is defined to be the number of sites at which the strings it connects differ. The objective is to find a phylogeny which spans a given set of species and which is of minimum total weight as a sum of the weights of its arcs. This aim is known in biology by the rather contradictory title of *maximum parsimony*.

It is not assumed that the evolutionary history of the given species necessarily followed the path laid out by the phylogeny of maximum parsimony. This tree is merely a minimal solution to an extremal problem in this model, a criterion which is often used to describe natural phenomena. This problem was defined in Section 9.3, where its complexity is analysed. Enumeration of the number of feasible solutions to it is discussed in Section 3.5.2.

We now turn to using this problem to test the theory of evolution. To observe the appearance, through evolution, of new species would require millions of years, (far beyond the compass of even the most far-sighted research organization). Consequently, evolution is mainly a retrospective study, concerned with evidence that evolution has actually happened. As such, the theory has been criticized for not being directly testable in the same way as, for example, Rutherford's model of the atom.

Let us now see what can be done about testing the theory. It has long been considered that protein sequence data may contain evolutionary information (Zuckerkandl and Pauling (1965).) In particular, the phylogeny of maximum parsimony makes no assumptions about the mechanism of evolution and has been widely used as a model for evolutionary relationships (Dayhoff and Eck (1972)). Given any comparative data, irrespective of origin, one can construct trees of evolutionary form, and hence a phylogeny of maximum parsimony must exist. So in general, finding such a phylogeny is not, in itself, independent evidence for the existence of an evolutionary tree.

However another prediction can be made from the theory of evolution. The prediction is that the maximum parsimony phylogenies spanning the same species, constructed from independent proteins should have a similar evolutionary history. We need a measure of tree similarity with biological significance so that the output can be compared with the hypothesis that the phylogenies produced are randomly selected. We describe a suitable measure later.

Our approach is to take different protein sequences for a common set of species, find all the phylogenies of maximum parsimony and those which are close to maximum parsimony and compare them. If the probability is high that these phylogenies are unrelated then this indicates that the protein sequences do not contain similar evolutionary information and hence would contradict the conjecture that there exists an evolutionary tree for those species.

In the next section we describe the first step in the testing procedure - the construction of a model of the problem of identifying a phylogeny of maximum parsimony.

15.3.3 A model for phylogeny construction

Our aim is to identify a phylogeny of maximum parsimony in which each given species and its sequence is represented. It is usual first to construct an undirected phylogeny (a spanning tree); which does not have a distinguished vertex representing the common ancestor of all the given species. A common ancestor is then specified by directing the phylogeny, that is, by giving each edge an orientation away from the common ancestral vertex.

We shall confine ourselves to the production of undirected phylogenies because our measure of comparison of trees is unaffected by whether or not the phylogenies are directed.

15.3.4 Solution of the Model

We now outline a heuristic method for the *Steiner Problem in Phylogeny* which we established is NP-hard in Section 9.3. It is based on a decomposition technique which employs a modified minimal spanning tree algorithm (see Section 12.5.1). To make this clearer, we now furnish a small artificial example which has been constructed for expository purposes. Suppose we have five species: $S_1, S_2, \ldots, S_5$, with strings that are only eight characters long rather that the usual 312 characters. The strings are given in Table 15.1, where each row is the string for a species.

	1	2	3	4	5	6	7	8
S_1	A	G	U	G	U	U	A	A
S_2	C	A	A	G	U	U	A	A
S_3	C	G	U	C	C	G	A	A
S_4	A	A	U	G	U	U	C	A
S_5	C	G	U	G	U	U	C	U

Table 15.1

What we want to do is to connect the five species by a spanning tree where each species is represented by a vertex of the tree. Suppose we join S_1 and S_4 directly by an edge in the tree, as shown in Figure 15.4. If we compare S_1 and S_4, we see that the two strings differ at columns 2 and 7 only. At column 2, S_1 has a G and S_4 has an A. We label the $S_1 S_4$ edge with the symbol $2GA$ to denote one difference between the strings. These three symbols together are called a *substitution*. We also add the substitution $7AC$ to the edge to denote the difference at column 7. As there are no other differences, there are no other substitutions associated with the $S_1 S_4$ edge. Given one of the two strings and the substitutions, we can deduce the other string. What we wish to do is to construct a spanning tree for $(S_1, S_2, S_3, S_4, S_5)$ and label the edges of the tree with the appropriate substitutions. A possible spanning tree is given in Figure 15.5. The total number of substitutions in this tree is 12. The objective is to find a spanning tree which minimizes the total numbers of substitutions.

Finding the phylogeny of maximum parsimony for the species in a realistic study is no easy task. As the number of possibilities grows factorially, complete enumeration is out of the question. The model appears on the surface

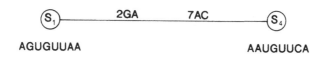

AGUGUUAA   AAUGUUCA

Figure 15.4 The (S₁,S₄) edge

to be a minimal spanning tree problem which is, of course, straightforward to solve by the algorithms of Section 12.5.1. However, the introduction of additional sequences representing extra species (either existing or extinct) often leads to a phylogeny with greater parsimony (a tree of less weight). Therefore, the model can be seen to be a variation of the Steiner tree problem, as explained in Section 9.3. It has been possible to modify existing minimal spanning tree algorithms to produce a phylogeny of relatively low weight. This phylogeny of low weight is then analyzed with view to either proving that it is of maximum parsimony, or modifying it so that it becomes one of maximum parsimony. The approach is to use the clustering algorithm to partition the matrix of data columnwise. This breaks up each original string into a number of smaller substrings. Each set of substrings represents a separate, smaller problem. These smaller problems are then analysed and if necessary a further decomposition is made. Eventually then each subproblem can be solved directly. The weights of their solutions provide valid lower bounds which are used to prove the minimality of either the original phylogeny or a modification.

We now illustrate the ideas just presented by returning to the example problem. A natural first attempt at solution is to use an algorithm for the minimal spanning tree problem (MSTP). See Section 12.5.1. Let us see how this approach turns out for our example. The number of substitutions necessary for each pair of strings is given in Table 15.2

Applying an MSTP algorithm we can obtain (among other minimal trees) the tree given in Figure 15.5. There is the possibility of the introduction of new strings in order to reduce the total number of substitutions. This is

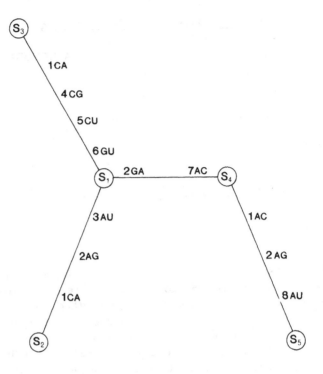

Figure 15.5 A spanning tree for the five species

	S_2	S_3	S_4	S_5
S_1	3	4	2	3
S_2		5	3	4
S_3			6	5
S_4				3

Table 15.2

brought about by using a process called *coalescement*. This process can be explained as follows. Suppose we have a substitution at column n between nucleotides X and Y, which appears on two adjacent edges in a spanning tree, as shown in Figure 15.6. The tree can be modified to produce a spanning tree with one less substitution by the introduction of a new string $\overline{S}_m$, which differs from the S_i string only at site n. As defined in Section 9.3, $\overline{S}_m$ is called a *Steiner vertex*. It has a Y rather than an X there. This process can be used repeatedly to reduce that number of substitutions. In applying it to our tree we can remove one instance of duplication $2GA$ (on edges $\{S_1, S_4\}$ and $\{S_4, S_5\}$) and of $1CA$ (on edges $\{S_3, S_1\}$ and $\{S_1, S_2\}$). This produces the tree shown in Figure 15.7 which has 10 substitutions. We have introduced two new strings: $\overline{S}_6$ and $\overline{S}_7$.

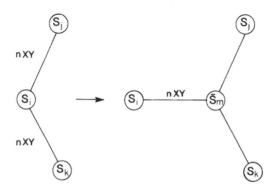

Figure 15.6   The coalescement problem

To discover whether this tree is a minimal Steiner tree, we analyse Table 15.1. Consider the three sites: 1, 2, and 7. If substitutions corresponding to the other sites are removed, the tree collapses to the one in Figure 15.8. It has one Steiner vertex. The introduction of more than one Steiner vertex would lead to a tree with more substitutions. Can we create a tree with fewer substitutions by having no Steiner vertices? This is a normal MSTP. The substrings based on sites 1, 2, and 7 are given in Table 15.3. Any minimal solution to the MSTP for this data has five substitutions.

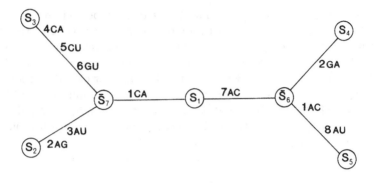

Figure 15.7 The Minimal Phylogeny

Therefore, we know that sites 1, 2, and 7 jointly require at least five substitutions in any phylogeny. The other sites 3, 4, 5, 6, and 8 require at least one substitution each, as their columns in Table 15.1 each have two nucleotides. Thus, any phylogeny must have at least $5+1+1+1+1+1=10$ substitutions. As the phylogeny constructed in Figure 15.7 has exactly 10 substitutions, it must be minimal. It is not uniquely minimal.

We now return to general, large problems. Sometimes the site decomposition process produces subproblems which are small enough to be solved by a certain DFS algorithm (see Section 9.5.1). This DFS algorithm guarantees to produce all phylogenies of maximum parsimony on any data set with no more than 12 species in reasonable computing time. So it was used to calculate directly the phylogenies for the 11 species problems mentioned earlier.

There is no guarantee that the decomposition technique just described will converge to a phylogeny of maximum parsimony when applied to a large data set.

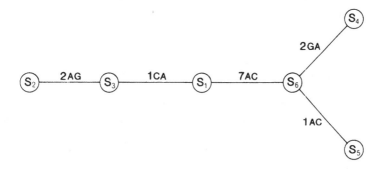

Figure 15.8 A phylogeny for sites; 1,2, and 7

	1	2	7
S_1	A	G	A
S_2	C	A	A
S_3	C	G	A
S_4	A	A	C
S_5	C	G	C

Table 15.3

15.3.5 Validation and implementation

We turn now to applying the heuristic method outlined in the last section to testing a specific prediction of the theory of evolution. The prediction is that similar phylogenies should be obtained from different protein sequences. The first task is to introduce a biologically meaningful measure of the difference between two phylogenies.

One of the main reasons why phylogenies are constructed is to help in the attempt to classify species into classes at different levels. Classes of related species are naturally formed by the descendents of a common ancestor. In a binary phylogeny on n species, there will be $(n - 2)$ such classes. We define the *difference* $d(P_1, P_2)$, between two phylogenies P_1 and P_2, to be the number of classes present in exactly one of P_1 and P_2.

This measure of phylogeny difference is a metric on the space of all phylogenies spanning the given n species. In any undirected phylogeny P, the removal of any internal edge will partition the set of species into two subsets, at least one of which will be a class of related species. The set of all such partitions uniquely defines P. Therefore $d(P_1, P_2)$ is the number of partitions of the species formed by suppressing internal edges which differ in P_1 and P_2. There are

$$(2n - 5)!! = (2n - 5) \times (2n - 7) \times (2n - 9) \times \ldots \times 5 \times 3,$$

different phylogenies on n species. We can for any specific phylogeny P, among these, determine $d(P, P')$ for each of the other phyolgenies P'. The proportion of phyolgenies with $d(P, P') = M$, represents the probability that a randomly selected phylogeny P', is distance M from P. As we are going to compare phylogenies, we need the probability of occurrence $d(P, P')$ values over all pairs of randomly selected phylogenies. These probabilities are given in Table 15.4. Some explanation of this table is in order. For binary phylogenies on n species, there are $(n - 3)$ internal edges. So each phylogeny possesses $(n - 3)$ partitions. If P_1 and P_2 have q partitions in common then $d(P_1, P_2) = 2(n - 3 - q)$, which is even. Further, $d(P_1, P_2)$ can range from 0 to $(2n - 6)$. The value E is the weighted average or expected value of $d(P_1, P_2)$, for two randomly selected phylogenies P_1 and P_2 on 11 species. Thus

$$E = \sum_{\substack{M=0 \\ M \text{ even}}}^{16} M f(M),$$

where $f(M)$ is the frequency of occurrence of $d(P_1, P_2) = M$.

There are five protein sequences available that are common to the 11 species that are used for the test of evolution discussed earlier. They are: cytochrome c, haemoglobin A and B, and fibrinopeptide A and B. The original data were taken from Dayhoff and Eck (1972). The sequences have been converted to nucleotide sequences (where the difference between each pair of symbols is exactly 0 or 1) and have been edited to remove invariant sites and other sites containing no comparative information. The final data is presented by Penny et al. (1982).

For 11 species there are 17!! (34.4 million) possible binary phyogenies. The methods discussed in the previous section have been applied to produce all minimal phylogenies. Table 15.5 displays the numbers of trees either minimal or close to minimal length, for each of the five protein sequence sets

$$M$$

0	2.9×10^{-8}
2	4.6×10^{-7}
4	4.7×10^{-6}
6	4.0×10^{-5}
8	3.3×10^{-4}
10	2.8×10^{-3}
12	2.3×10^{-2}
14	1.7×10^{-1}
16	8.1×10^{-1}
E	15.55

Table 15.4 The frequency of occurrence of
$d(P_1, P_2) = M$ for binary phylogenies on 11 species.

and for an artificial sequence produced for each species by concatenating all five of its sequences. There are 39 phylogenies within 1.25% of optimality. There were two identical trees among the 39. Using the comparison technique outlined previously, we obtain $\binom{39}{2}$ ($= 741$) values of $d(P_1, P_2)$, ranging from 0 to 14, out of the maximum of 16, with a mean value of 7.57. Interpolating from Table 15.4, we expect such a value to occur for a pair of randomly selected phylogenies with probability 1.9×10^{-4}. There are 741 comparisons to be made but they are not all independent. The mean similarity for comparisons between phylogenies from the same sequence is 4.1 and for different sequences is 8.33. On looking at Table 15.5, we can see that there is a strong divergence away from random selection towards the phylogenies being much more similar than one would expect from random phenomena.

M	0	2	4	6	8	10	12	14	16
Expected	0	0	0	0	0	2	17	125	597
Observed	1	53	87	163	200	145	84	8	0

Table 15.5

Table 15.6 gives the average values of $d(P_1, P_2)$ between phylogenies of different sequences. The upper values are obtained from only minimal phylogenies and the lower values from all 39 phylogenies. The largest average is 11.75, between the minimal phylogenies of haemoglobin B and fibrinopeptide B. This value occurs between a pair of random trees with probability 1.8×10^{-2}. All other values have probabilities less than this, down to 1.1×10^{-5} for fibrinopeptides A and B.

It is obvious that the phylogenies from different sequences are not independent. The different protein sequences produce phylogenies which are remarkably similar and display relationships between them consistent with the prediction from the theory of evolution. This provides evidence to support the theory but does not, of course, prove it.

	CS	Cc	FA	FB	HA	HB
CS	2.0	7.0	8.0	9.8	8.0	5.3
Cc	5.8	3.1	8.3	10.2	6.0	5.7
FA	6.9	8.3	–	4.8	10.0	11.3
FB	8.2	10.2	4.8	4.5	8.8	11.8
HA	8.4	7.6	10.3	11.2	4.7	6.7
HB	6.5	7.6	11.1	11.5	9.0	2.7

Table 15.6 The average distance between trees derived from two sequences.

15.4 Summary

In this chapter we have discussed applications of graph theory in physics, chemistry, and biology. More specifically we have considered, from a graph theoretic point of view, the dimer problem in physics, molecular representation and isomer enumeration in chemistry, genetic structure and evolutionary trees in biology. We have seen that graphs have been used in science for two centuries. Initially graphs were used as diagrams only, to classify existing knowledge. More recently graphs have been used to gain new insights in science via deeper results from the mathematical theory. These insights have been gained in areas including: statistical continuum mechanics, bonding theory, chemical kinetics, polymer systems, genetics, and evolution.

Thus there has been a useful two-way process, with the generation of: (i) new results in graph theory significant in their own right, triggered by practical problems in science, and (ii) the breaking of new ground in science with the application of these graph theoretic results.

15.5 Exercises

15.1 Display all dimer coverings of the $n \times n$ rectangular lattice for $n = 2$, 3, and 4.

15.2 Draw a graph, other than the one shown in Figure 15.1(b), which has an even number of vertices, but no dimer coverings.

15.3 Draw the two isomers of butane, the three isomers of pentane, and the five isomers of hexane.

15.4*Show that an n-cycle cannot be represented by an interval graph unless $n = 3$.

15.5*Prove Theorem 15.1. Hint: see Gilmore and Hoffman (1964).

15.6 Construct a minimal phylogeny for the following data. Hint : The optimal phylogenies have 12 substitutions and one Steiner vertex.

$$
\begin{array}{c|cccccc}
S_1 & C & U & U & G & A & G \\
S_2 & G & U & C & C & A & G \\
S_3 & G & U & C & C & A & A \\
S_4 & C & G & C & C & C & C \\
S_5 & C & G & U & C & C & U \\
S_6 & A & G & C & U & A & G
\end{array}
$$

15.7 Find a minimal spanning tree for the above data, which does not contain Steiner vertices.

15.8 Calculate the percentage saving in the total number of substitutions in the phylogeny identified in Exercise 15.6, compared with the phylogeny identified in Exercise 15.7.

15.9*Show that for any phylogenetic data the length of the minimal phylogeny is no less that 50% of the length of the minimal spanning tree, in which Steiner vertices are not allowed. Hint: see Foulds (1984).

16 Civil Engineering

The two hour's traffic of our stage.

Romeo and Juliet, Prologue, William Shakespeare

In this final chapter we discuss the application of graph theory in civil engineering. We begin with a very brief description of a problem concerning the construction of earthwork projects. We then go on to the major application of this chapter, which involves the design of urban traffic networks. It represents a major application of digraphs and networks. Further applications of graph theory in civil engineering, involving highway route and pier construction planning, are outlined by Chachra et al. (1979).

16.1 Earthwork projects (Based on Section 6.2, Chachra et al. (1979).)

There are many civil engineering projects which involve the construction of large earthworks, such as canals, raised roads or railways, certain buildings, and dams. The earth filling required for such projects usually comes from *borrow pits,* which are areas of suitable ground near the project site from which the filling is dug. This filling must be transported from the borrow pits to various areas of the project site. Each pit has a known amount of filling available and each area has a known required amount of filling. The

343

problem is to transport the filling at minimal total cost with the proviso that availability at each pit is not exceeded and the requirement at each area is met.

This problem can be modelled as a bipartite graph $G = (V, E)$ with vertex partition V_1 and V_2, where $V_1 \cup V_2 = V$. Here V_1 and V_2 represent the pits and the areas respectively. An edge $v_1 v_2$, with $v_1 \in V_1$ and $v_2 \in V_2$ is present in G if and only if it is possible to transport filling from pit v_1, to area v_2. Each pit represented by a vertex in V_1 has associated with it the known amount of its available filling. Likewise, each area represented by a vertex in V_2 has associated with it the known requirement of each area. Each edge $v_1 v_2$, of E has associated with it the unit cost of transporting filling from pit v_1 to area v_2. An instance of this problem is shown in Figure 16.1. Note that the unit costs incurred in transporting filling to area 8 are zero. Zero unit costs are assigned to a dummy pit (or area) when the sum of the availabilities does not equal the sum of the requirements. The dummy pit (area) thus introduced is assigned an availability (requirement) which causes the above-mentioned sums to be equal. In the above case, the original problem has pits 1, 2, and 3, with a total availability of 105 units. It has areas: 4, 5, 6, and 7, with a total requirement 95 units. Thus a dummy area 8, is introduced with a requirement of 10 units and zero unit costs.

This problem is an example of the transportation problem which is discussed, along with algorithms for its solution, in Sections 12.5.3 and 12.6.1.

16.2 Traffic Network Design (Based on Foulds (1985).)

One of the most important trends in transportation is the accelerating pace of motorized transport. The growth of vehicle fleets on a world-wide basis is of the order of 15 percent per annum. Almost every country in the world is experiencing such growth. The rapid trend toward urbanization throughout the world, coupled with the growth of motorization, is creating problems of urban transport on a wholly new scale for both developed and less developed countries.

Transportation has significant effects on where and how cities grow and how they function. The convergence of transport routes, or the transshipment points from one mode to another, frequently provide the focus and impetus for urban development. Mobility and accessibility to goods, labour, and markets permit and encourage the urban concentration. Development patterns within cities are, to a large extent, shaped by transportation investment decisions which have a large impact on land values and land uses, just as the latter have an important bearing on transport requirement. Among other important reasons why transportation consistently ranks among the

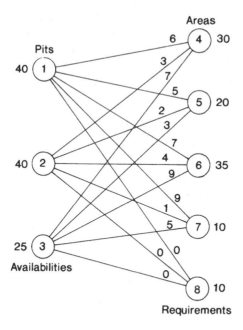

Figure 16.1 A graph theoretic representation
of an earthwork project

chief concerns of cities is that it is a prerequisite for jobs and business, it accounts for a substantial part of personal and public expenditure, and it typically occupies from 20 to 40 percent of a city's physical space.

Transportation flow problems concerned with a network of roads arise in many different practical contexts. Government planners at various levels (city, regional, and national) often find it necessary to attempt to change traffic patterns in existing networks. Changes may be made by altering an existing network in many ways, including: arc addition or deletion, capacity increase of certain arcs, time reduction in traversing certain arcs, flow reduction in congested sectors by the introduction of traffic control or mass transit systems, or the re-orientation of the direction of flow in certain arcs.

Consideration must be given to the recognition of trends. Most problems change over time, and some thought should be given to forecasting future data. This can be used to anticipate and plan for a problem that is likely to become a reality at some later date. Future volume can be forecast by correlating past and present volume for areas, zones, and subzones to such indices as past and present population, land use, vehicle registration, fuel consumption, industrial production, and other economic factors.

Securing an accurate description of the existing network and its flow pattern is the first step towards solving any given problem. To obtain the necessary data, the following questions, among others, should be asked.

(i) What are the magnitudes of flow volumes? Are these likely to change? How? Is the flow homogeneous?

(ii) Where does the traffic originate? Where are the sources or generators of the traffic?

(iii) Is it possible to introduce new modes of traffic flow to increase link capacity? (One such mode is a rapid transit system.)

(iv) Where is the traffic going? Where are the major final destinations of the flow?

(v) What are the current capacities of the network links? Can they be increased? How?

(vi) At what speed is the traffic moving? What are the estimates of times taken or distances travelled to traverse the links in the network? Are these estimates dependent upon the levels of flow in the links? How? Can these estimates be made more attractive by improving certain links?

(vii) What energy levels are used in crossing the links of the network?

16.2.1 Simple traffic network design models

We begin this section with a simple traffic network design model. It concerns an urban traffic network of two-way streets. Suppose that it is believed that the traffic in the network would flow more efficiently if each street is declared one-way. The question is how to choose the orientation of each street so that it is still possible to travel from any intersection to any other intersection in the network. We shall consider more complicated design models, including those concerning one-way street declaration, later in this chapter.

We can model the above-mentioned problem by a graph which has vertices and edges which represent intersections and (two-way) streets, respectively. Then the original problem is equivalent to orienting the graph so that the resulting digraph is strongly connected. (This concept, along with many others for digraphs is defined in Section 7.1) Recall from Section 1.2, that a graph is termed *oriented* if each of its edges is given an orientation, or direction and thus turned into an arc.

Of course, not every graph has a strongly connected orientation. Certainly, disconnected graphs have not. However, certain connected graphs do not possess strongly connected orientations, for instance, the connected graphs in Figure 16.2. These graphs all have at least one *bridge*, as defined in Section 2.1 and Exercise 9.8. Indeed the possession of no bridges is an obvious necessary condition for a graph to have a strongly connected orientation. The following theorem shows that this condition is also sufficient.

Theorem 16.1

A connected graph G does not have any bridges if and only if G has a strongly connected orientation.

Proof

The reader is referred to the proof given by Robbins (1939).

It is possible to find a strongly connected orientation in any connected bridge-less graph G, by applying the DFS approach to G, as introduced in Section 9.5.1. Each edge of G is given the orientation assigned to it as the DFS algorithm is employed.

This orientation process for the graph in Figure 9.3(a) is illustrated in Figure 9.4(a), where a strongly connected orientation results. Roberts (1976) has shown that, as long as G is connected and without bridges, that DFS will guarantee a strongly connected orientation. Naturally, the converse of the digraph created by the DFS orientation is also strongly connected. This algorithm has complexity $O(n^2)$, where n is the number of vertices in G.

A standard design problem will be introduced next. It involves choosing additions or improvements to an existing network in order to reduce traffic congestion. Decisions about such choices often become necessary when usage of the network approaches levels beyond those for which it was designed. In order to ease the resulting congestion, the capacity of the network must be expanded. The problem can be modelled as follows. The zones or major street intersections and the major streets or freeways, of an urban area can be represented by the nodes and arcs respectively, of a network. It is usual to assume Wardrop's second principle (1952) of driver behaviour in predicting the arc flow levels. This principle is that, for every origin-destination pair of nodes p_i and p_j:

(a) The travel cost to each driver must be the same on all $p_i - p_j$ routes actually used.

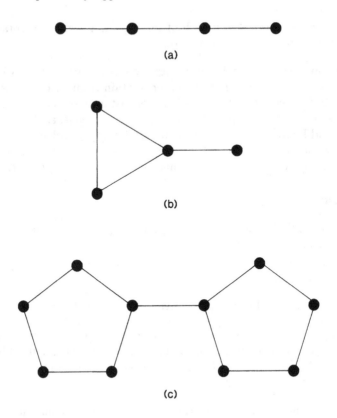

Figure 16.2 Connected graphs which do not possess strongly
connected orientations

(b) There is no unused $p_i - p_j$ route with a smaller travel cost than that
of those actually used.

Arc flow levels which are predicted while assuming the above are said to
be *user-optimal* or *equilibrium flows*.

Beckmann et al. (1956) have shown that the equilibrium flows are given
by the solution to the following problem.

Problem 1 (P1) [Predicting Equilibrium Flows]

$$\text{Minimize} \quad \sum_{(i,j)\in A} \int_o^{x_{ij}} A_{ij}(t)dt \qquad (16.1)$$

subject to

$$D_{js} + \sum_{(i,j)\in A} x_{ij}^s = \sum_{(j,k)\in A} x_{jk}^s, \; j=1,2,\ldots,n, \; s=1,2,\ldots,n, \; s\neq j \; (16.2)$$

$$\sum_{s=1}^n x_i^s = x_{ij}, \quad (i,j)\in A, \qquad\qquad\qquad (16.3)$$

$$x_{ij} \geq 0, \quad (i,j)\in A, \qquad s=1,2,\ldots,n. \qquad (16.4)$$

where:

$A =$ the set of network arcs,

$n =$ the number of nodes in the network,

$D =$ the number of drivers with origin node j and destination node s,

$x_{ij} =$ the number of drivers who use arc (i,j),

$x_{ij}^s =$ the number of drivers with ultimate destination s who use arc (i,j) and,

$A_{ij}(x_{ij}) =$ the average user cost of arc (i,j) when it has x_{ij} users.

A commonly-used estimate of $A_{ij}(x_{ij})$ is $a_{ij} + b_{ij}x_{ij}^4$. Here a_{ij} is a free running cost and b_{ij} is a congestion parameter.

LeBlanc (1975) has shown that each optimal x_{ij} for (P1) is unique, if each function A_{ij} is increasing in x_{ij}. The solution to Problem (P1) is used to identify which improvements should be made to the network in order to maintain adequate service as usage increases and road surfaces deteriorate. The objective is to invest funds subject to a budget constraint so as to minimize total congestion. The problem can be modelled as follows:

Problem (P2) (The Traffic Network Design Problem)

$$\text{Minimize} \quad \sum_{(i,j)\in A} A_{ij}(x_{ij})x_{ij}, \qquad (16.5)$$

subject to

$$\sum_{i=1}^n \sum_{j=1}^n c_{ij}y_{ij} \leq B, \qquad\quad i\neq j \qquad (16.6)$$

$$x_{ij} \leq My_{ij}, \text{ for each potential arc } (i,j), \qquad (16.7)$$

and

$$X \in \overline{X} \qquad\qquad (16.8)$$

where

    $c =$ the cost of improving arc (i, j), if it already exists, or of constructing it, if it does not. The parameter is defined over all $p_i - p_j$ pairs. The cost c_{ij} may be assigned a prohibitively large value if there is no possibility of constructing the potential arc (i, j),

    $B =$ the maximum amount of funds that can be invested,

    $y_{ij} = 1$, if arc (i, j) is improved or constructed,

       $= 0$, otherwise,

    $\overline{X} =$ the set of optimal solutions to Problem (P1), and

    $M =$ a suitably large number.

The model can be interpreted as follows. (16.5) indicates the total user cost of all drivers, which is taken as the measure of congestion, to be minimized. (16.6) is the budgetary constraint. (16.7) prevents flow in a potential arc which is not constructed. (16.8) ensures that the flows are user-optimal. Note that Problem (P2) is not a normal optimization problem because of (16.8), which depends upon Problem (P1). The only decision variables in Problem (P2) are the y_{ij}'s.

We turn now to developing a network design model which is significantly different from Problem (P2).

16.2.2 An Arc Elimination Model

When a network is uncongested, travel costs bear little relation to flow levels. In this case the addition of a new arc will not increase any driver's cost and hence will not increase total congestion. Intuitively, it seems likely that this will also be true when congestion is present and travel costs are sensitive to flow levels. Such is not always the case. Braess (1968) constructed a numerical example in which the addition of a new arc significantly increased all driver's costs and therefore increased congestion. This result was made more accessible by Murchland (1970) who dubbed it *Braess's Paradox* and commented on its implications. A slightly more elaborate example is given by LeBlanc (1975). This result is not really paradoxical in the sense that there is no scientific reason why arc addition should lessen congestion, given user-optimal flows. The papers referred to so far have considered only small, contrived examples. However Knodel (1969) reported that Braess's paradox may have occurred in actual practice in the city of Stuttgart. After network improvements failed to achieve the desired results, an arc (representing the lower part of Königstrasse) was eliminated. This resulted in a reduction in congestion.

Traffic engineers often use arc elimination as a strategy to ease congestion in urban traffic networks. Arc elimination may increase the capacity of a network to accommodate flow for a variety of reasons. Arcs usually occur in oppositely directed pairs. When exactly one arc of a pair is eliminated, the street the pair represents becomes one-way. This may increase network capacity because:

(i) Drivers no longer wait at the intersection of one end of the street because turning into the street is now prohibited.

(ii) Drivers may drive faster in the street as there is no oncoming traffic.

(iii) A multi-laned street can usually handle more flow if all lanes have the same direction. There need not be more than one under-utilized lane whose direction is opposite to that of the majority of users.

Arc elimination in a network can be used to reflect the prohibition of turns in the street scene it models. This is done in the METRA computer package (1964) by representing each intersection of four streets by a subnetwork of eight nodes and 20 arcs, as in Figure 16.3. The elimination of certain of the arcs corresponds to certain turn prohibitions. This technique is explained in the book by Potts and Oliver (1972).

We now construct a model for the problem of which arcs to eliminate in order to ease congestion. It is assumed that:

(i) There is a known deterministic demand between each origin-destination pair.

(ii) Arc traversal cost is non-linearly dependent upon flow.

(iii) Flows are user-optimal. That is, the flows obey Wardrop's second principle.

It is further assumed in the model to be developed that each arc (i, j) has an absolute capacity u_{ij}, which cannot be exceeded when arc (j, i), is originally present. If arc (j, i) is eliminated, the capacity of (i, j) is increased to $u_{ij} + u_{ji}$. The motivation for this is that with the elimination of (j, i) the street connecting intersections i and j becomes one-way in the i-to-j direction. The half of the street formerly used for flow in the j-to-i direction is now also available for i-to-j flow and thus u_{ij} is increased by u_{ji}. There are instances when exactly one of the arcs (i, j) and (j, i) are present in the original network. This will occur, for example, in the case of a one-way street and when an intersection is represented by a subnetwork of arcs for the purposes of turn prohibition as in the METRA package. Naturally u_{ij} is set equal to zero for each instance in which arc (i, j) is absent. This means that there is no increase in capacity in an arc when a nonexistent opposing arc is *eliminated*.

The problem of which arcs to eliminate in order to minimize congestion can be modelled as

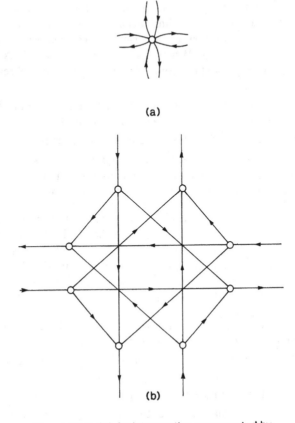

(a)

(b)

Figure 16.3 (a) An intersection represented by
(b) a subnetwork

Problem (P3) (The Arc Elimination Problem)

$$\text{Minimize} \quad \sum_{(i,j)\in A} A_{ij}(x_{ij})x_{ij}, \tag{16.9}$$

subject to

$$x_{ij} \leq u_{ij} + (1 - y_{ij})u_{ji}, \qquad (i,j) \in A \tag{16.10}$$

$$x_{ij} \leq (u_{ij} + u_{ji})y_{ij}, \qquad (i,j) \in A \tag{16.11}$$

$X \in \overline{X}$, the set of user-optimal flows for the modified network, (16.12)

where:

$$y_{ij} = 0 \text{ if arc } (i,j) \text{ is eliminated,}$$
$$1 \text{ otherwise, and}$$
$$u_{ij} = \text{capacity of arc } (i,j).$$

The model can be interpreted as follows. Total congestion is minimized as reflected in (16.9). (16.10) increases the capacity of an arc (i,j) from u_{ij} to $u_{ij} + u_{ji}$ if its partner (j,i) is eliminated, and vice versa. (16.11) prevents flow in an eliminated arc. (16.12) ensures that flows are user-optimal. Because of (16.12) the model is not a straight-forward mathematical programming problem but is dependent upon Problem (P1).

16.2.3 An algorithm for problem (P3)

The algorithm to be introduced is a BFS branch and bound approach. The general principles of branch and bound were discussed in Section 14.2.3. An algorithm which applies the branch and bound method to Problem (P3) is described next. A solution to Problem (P3) is given by specifying values for the variables:

$$y_{ij}, \qquad i = 1, 2, \ldots, n; \quad j = 1, 2, \ldots, n; \quad i \neq j.$$

They can be assembled to form a vector:

$$y = (y_{12}, y_{13}, y_{14}, \ldots, y_{1n}, y_{21}, y_{23}, y_{24}, \ldots, y_{2n}, \ldots, y_{n1}, y_{n2}, \ldots, y_{nn-1}).$$

A partial solution can be represented by displaying the values of the variables decided upon and leaving the other entries blank. Thus the decision to eliminate (1,2) and (1,3), leave (1,4) intact, and to leave the fate of the other arcs undecided upon as yet, can be represented by $(0, 0, 1, _-, _-, _-, \ldots, _-)$. As each node z, in the tree corresponds to a partial solution, we can associate three sets with it:

$$E(z) = \{(i,j) : y_{ij} = 0\}, \qquad (\textbf{E} \text{ for eliminate}),$$
$$I(z) = \{(i,j) : y_{ij} = 1\}, \qquad (\textbf{I} \text{ for intact}), \quad \text{and}$$
$$F(z) = \{(i,j) : y_{ij} \text{ undecided}\}, \qquad (\textbf{F} \text{ for free}).$$

In the previous example:

$$E(z) = \{(1,2), (1,3)\}, \quad I(z) = \{(1,4)\}, \quad \text{and} \quad F(z) = A\backslash(E(z) \cup I(z)).$$

The set of successors $S(z)$, of a partial solution $P(z)$, corresponding to a node z is defined as the set of all solutions containing $P(z)$, i.e corresponding to z. That is,

$$S(z) = \{Y : y_{ij} = 0, (i,j) \in E(z), y_{ij} = 1, (i,j) \in I(z)\}.$$

In the example:

$$S(z) = \{(0,0,1,1,1,\ldots,1), (0,0,1,1,1,\ldots,0), \ldots, (0,0,1,0,0,\ldots,0)\}.$$

At each iteration a node z is selected and, if necessary, the partitioning routine partitions $S(z)$ into two disjoint subsets by sprouting two nodes z_1 and z_2 ,say, both emanating from z.

E, I, and F for z_1 are defined by selecting an element of $F(z)$, (p,q) say, where:

$$E(z_1) = E(z) \cup \{(p,q)\}, \quad I(z_1) = I(z), \quad F(z_1) = F(z_2) = F(z)\backslash\{(p,q)\}$$
$$E(z_2) = E(z), and \quad I(z_2) = I(z) \cup \{(p,q)\}.$$

The kernel of the method is the calculation for each new node z, in the tree ,of a lower bound, denoted by $B(z)$, on the value of any solution in $S(z)$. Let $T(z)$ be the value of the optimal solution to the problem generated by adding the following constraint to Problem (P3).

$$Y \in S(z). \tag{16.13}$$

We denote this problem by **(P4)**.

Then we wish to find $B(z)$, where

$$B(z) \leq T(z).$$

Intuitively, one would expect to be able to calculate a reasonable estimate of $B(z)$ by using the optimal solution to the problem generated by adding the following family of constraints to (P3):

$$\begin{aligned} y_{ij} = &0, \quad (i,j) \in E(z), \\ &1, \quad (i,j) \in I(z) \cup F(z). \end{aligned} \tag{16.14}$$

In this case there is no optimizing to be done, as all the decision variables, (the elements of Y) have predefined values. The problem reduces to finding the user-optimal flow, and hence total congestion, when all free arcs are left intact. Unfortunately, as LeBlanc (1975) points out, because of Braess's paradox, this strategy may not produce a valid lower bound. Instead we calculate $B(z)$ as follows. Consider a problem which is identical to (P4) except that flows which minimize (16.9), rather than user-optimal flows, are assumed and there are no arc capacity constraints:

Problem (P5)

$$\text{Minimize} \quad \sum_{(i,j)\in A} A_{ij}(x_{ij})x_{ij}, \tag{16.15}$$

subject to (16.2), (16.3), (16.4) and (16.13).

That is, Problem (P5) differs from the previous models in that flows can be prescribed rather than having to be predicted. Let $T(z)$ be the value of the optimal solution to (P5). As a result of the following theorem, we can set $B(z) = T(z)$.

Theorem 16.2

For any node z in the decision tree,

$$T'(z) \le \hat{T}(z).$$

Proof

We begin by introducing a variation of Problem (P5) in which constraint (16.15) is replaced by (16.13). Let $\hat{T}(z)$ be the optimal solution value for this relaxed problem, which is denoted by (P6). Recall that $T(z)$ denotes an optimal solution to (P4). Any feasible solution of (P4) comprises a set of values for Y. However any feasible solution to (P6) comprises a set of values for both X and Y. (P4) and (P6) both possess the constraints: (16.2), (16.3) (16.4), and (16.13); (P4 having the additional constraints: (16.8), (16.10), and (16.11).) Clearly, if Y^* is an optimal solution (with value $T(z)$) for (P4) with corresponding flow matrix X^*, then (Y^*, X^*) is a feasible solution for (P6). Thus $\hat{T}(z) \le T(z)$.

Consider any optimal solution (P6). If it is such that $y_{ij} = 1$ for all $(i, j) \in F(z)$, it is feasible for (P5). Otherwise it can be made feasible by redefining as 1 the values of any y_{ij} which are currently 0 among all $(i, j) \in F(z)$. Thus any optimal solution for (P6) is feasible for (P5). Therefore the result follows. ∎

The implication of this theorem is that it provides a simple way to calculate the lower bound for any node z in the decision tree. Simply leave all free arcs intact and solve for the system-optimal flows. The resulting total congestion is a valid lower bound. As LeBlanc (1975) points out there is no question of the validity of this bound as Braess's paradox is concerned solely with user-optimal flows.

Suppose a node z is replaced by two new nodes: z_1 and z_2 where (i, j) is the arc of $F(z)$ upon which the branching is based. $B(z_2)$ can simply be set equal to $B(z)$ because the problems (P5) for z_1 and z are identical. This is because y_{ij} is permanently set equal to 1 in (P5) for z_2 and temporarily set equal to 1 in (P5) for z. Thus only one new bound $B(z_1)$, needs to be calculated in each partitioning routine. Arc (i, j) should be chosen so as to make $B(z_1)$ as large as possible. This is achieved by identifying the largest ratio x_{ij}/u_{ij}, among all arcs, say x_{pq}/u_{pq}, and choosing (i, j) to be (p, q). The fathoming procedure for node z_2 is based on the constraints (16.10) and (16.11). It ensures that they are satisfied by any feasible solution.

Consider an instance of problem (P5) which is to be solved to produce $B(z)$ for a particular node z. Let $A' = A \backslash E(z)$, the arc set left in the network for z.

Using the form for $A_{ij}(x_{ij})$ as defined earlier, the total user cost on arc (i, j) is

$$A_{ij}(x_{ij})x_{ij} = a_{ij}x_{ij} + b_{ij}x_{ij}^5.$$

(P5) then becomes:

Problem (P5′)

$$\text{Minimize} \quad \sum_{(i,j)\in A'} (a_{ij}x_{ij} + b_{ij}x_{ij}^5) \tag{16.16}$$

subject to

$$D_{js} + \sum_{(i,j)\in A'} x_{ij}^s = \sum_{(i,j)\in A'} x_{ij}^s, \qquad j = 1, 2, \ldots, n, \\ s = 1, 2, \ldots, n, \quad s \neq j \tag{16.17}$$

$$\sum_{s=1} x_{ij}^s = x_{ij}, \qquad (i, j) \in A' \tag{16.18}$$

$$x_{ij}^s \geq 0 \qquad (i, j) \in A' \quad s = 1, 2 \ldots, n. \tag{16.19}$$

LeBlanc has shown that the Frank-Wolfe algorithm (1956) is an effective algorithm for (P5). It is an iterative, globally convergent algorithm which, because of the structure of (P5′), requires only the repeated solution of shortest-path problems and one-dimensional searches. It is used to compute the lower bounds for the branch and bound procedure.

Algorithmic Statement of the Procedure

Notation

Let

$N = (V, A)$ be the network for the problem to be solved,

$n =$ the number of nodes in V,

$D_{js} =$ the given number of drivers with origin node j and destination node s,

$x_{ij} =$ the number of drivers currently using arc (i, j),

$A_{ij}(x_{ij}) =$ the average user cost of arc (i, j) when it has x_{ij} users,

$u_{ij} =$ the capacity of arc (i, j),

$y_{ij} = 0$, if arc (i, j) is eliminated,

$\quad = 1$, otherwise,

$Y = (y_{12}, y_{13}, \ldots, y_{1n}, y_{21}, y_{23}, \ldots, y_{2n}, \ldots, y_{n1}, y_{n2}, \ldots, y_{nn-1})$,

$E(z) = \{(i, j) : (i, j) \in A, y_{ij} = 0\}$,

$$I(z) = \{(i,j) : (i,j) \in A, y_{ij} = 1\},$$
$$F(z) = A\backslash(E(z) \cup I(z)),$$
$$S(z) = \{Y : y_{ij} = 0, (i,j) \in E(z); y_{ij} = 1, (i,j) \in I(z)\},$$
$Y(z) = $ the value of the optimal solution to the problem generated by adding the constraint $y \in S(z)$ to (P2), and
$$A' = A\backslash E(z).$$

Step

(0) (Initialization)

Set $r = 0$, $F(z_r) = A'$, $I(z_r) = \emptyset$, $E(z_r) = (N \times N)\backslash A$,

$y_{ij} = $ "-" if (i,j) is undecided upon

0, for all other i,j pairs, $i = 1, 2, \ldots, n$, $\quad j = 1, 2, \ldots, n$.

(1) (Initial flow assignment)

Assign all flow to N by solving (P5) by the Frank-Wolfe algorithm

(2) (Preliminary feasibility analysis)

If it is impossible to assign all flow because there is no path between at least one origin-destination pair j,s with $D_{js} > 0$, terminate the procedure as no feasible solution exists.

(3) (Identification of initial critical arcs)

Temporarily set $y_{ij} = 1$, $(i,j) \in F(z)$.

If constraint (16.10) is satisfied for user-optimal flow, terminate the procedure as the optimal network is N.

Otherwise continue.

(4) (The branching step)

For the arc (i,j) which does not satisfy (16.10) calculate

$$\alpha_{ij} = x_{ij}/u_{ij}$$

Let (p,q) be the arc with the maximum α_{ij} value. Create branch and bound decision tree nodes: z_r and z_{r+1} based on arc (p,q). Set

$$E(z_{r+1}) = E(z_r) \cup \{p, q\},$$
$$I(z_{r+1}) = I(z_r),$$
$$F(z_{r+1}) = F(z_r)\backslash\{(p,q)\},$$
$$F(z_{r+2}) = F(z_{r+1}),$$
$$E(z_{r+2}) = E(z_r), \qquad \text{and}$$
$$I(z_{r+2}) = I(z_r) \cup \{(p,q)\}.$$

(5) (Calculation of node bounds and node fathoming)

For each new node z in the decision tree calculate $B(z)$.

(6) (Fathoming)

Set $B(z) = \infty$ for any node z for which either (i) or (ii) is satisfied.

    (i) The criterion of step (2) is met.

    (ii) (a) At least one of the user-optimal, origin-destination paths, joining say j to s, for $D_{js} > 0$ is such that all its arcs are members of $I(z)$.

        (b) There exists an arc of this path (p, q) say such that $D_{js} > u_{pq} + u_{qp}$.

(7) (Test for infeasibility)

If all nodes have been eliminated from the decision tree terminate the procedure as no feasible solution exists.

Otherwise continue.

(8) (Identification of best node)

Identify the decision tree node z, say, with lowest $B(z)$ value among all active nodes. If the user-optimal flow assignment with $Y(z)$ satisfies constraints (16.10) and (16.11) terminate the procedure. ($E(z)$ specifies the optimal set of arcs to be eliminated.) Otherwise go to step (4).

A branch and bound technique for a variation of the network design problem has been discussed. The model involves arc capacity constraints, turn prohibitions, and the creation of one-way streets in order to minimize congestion. It was shown that the problem can be solved by solving a number of one-dimensional searches and shortest path problems. As both of these subproblems can be solved efficiently on large networks, it indicates that the technique can be used to solve realistically-sized problems in reasonable computational time.

16.3 Summary

In this chapter we have completed our exposition of graph theory and its applications. We have ended with graph theoretic applications in civil engineering. We began with a brief outline of a problem concerning earthwork projects. The major application of the chapter (and one of the major applications of this book) concerns traffic network design. We have showed how digraphs and networks can be used to construct and analyse a number of design models and also to provide various algorithms for their solution.

16.4 Exercises

16.1 Add one edge to each of the graphs in Figure 16.2, so that they no longer contain bridges. Apply DFS to each modified graph to produce a strongly connected orientation.

16.2*The structure resulting from removing the orientation of arc uv, whenever uv and vu are both arcs of a digraph D, is termed a *mixed graph* G. The arcs that are no longer oriented are called *edges*. G is termed *strongly connected* if D is strongly connected. G is termed *connected* if the multigraph, obtained by removing the orientation of all arcs in D, is connected. A *bridge* in a mixed graph is defined analogously to that in a graph.

Let G be a strongly connected mixed graph with edge $\{u, v\}$. Let D_1 be the digraph obtained from the underlying digraph D, of G by removing arcs (u, v) and (v, u) from D. Let $R_u^1 = \Gamma(v)\backslash\{u\}$, and $R_v^1 = \Gamma(v)\backslash\{v\}$. If $v \notin R^1$ and $u \notin R^1$, show that $\{u, v\}$ is a bridge of G.

16.3 Use the result proved in Exercise 16.2 to show that if G is a strongly connected mixed graph, then there is an orientation of every edge of G which is not a bridge which leaves the resulting mixed graph still strongly connected.

Further Reading

In this section we mention just a few of the many books available on the various topics discussed.

General

There are numerous books on graph theory which cover the topics of Part I of the present book. The first comprehensive treatment was written by König (1936). Since then many other advanced-level texts have been written, including Berge (1962), Harary (1969), Berge (1973), Bondy and Murty (1976), Behzad et al. (1979), Bollobás (1979), and Tutte (1984). The last-mentioned is by one of the leading exponents in the field and is a classic. Of the many intermediate level treatises, we mention those by: Kaufmann (1967), Wilson (1972), Deo (1974) (which contains many applications in science and engineering), Carré (1979) (which contains a major chapter on networks), and Tucker (1980) (which places graph theory in the wider context of combinatorics). We also recommend the elementary-level book by Trudeau (1976) and an interesting account of examples and counterexamples in graph theory by Capobianco and Molluzzo (1978).

Specialized Theoretical Topics

For more specialized theoretical topics, we point out the book by Biggs et al. (1977) on the history of graph theory from 1736 to 1936. Also worthy of mention is the text on graphical enumeration by Harary and Palmer (1973), and that on tree enumeration by Moon (1970). Treatises on the theory of matroids have been written by Von Randow (1975), Welsh (1976), and

White (1986) and on their applications by Lawler (1976), Recski (1989), and White (1992).

Digraphs

Some of the works with significant material on digraphs include: Harary et al. (1965) (with applications in the social sciences), Behzad et al. (1979), and at a lower level: Kaufmann (1972) and Robinson and Foulds (1980).

Networks

Busacker and Saaty (1965), Plane and McMillan (1971), Bazaraa and Jarvis (1977), Minieka (1978), Mandl (1979), Jensen and Barnes (1980), Kennington and Helgason (1980), Phillips and Garcia-Diaz (1981), Smith (1982), Tarjan (1983), and Aho (1983) contain significant material on networks and algorithms for them.

Algorithms

Christofides (1975), Lawler (1976), Chachra et al. (1979), Read (1979) and Gibbons (1985) have written specialized works on graph theoretic algorithms. The following devote considerable attention to the topic: Deo (1974), Minieka (1978), Carré (1979), Tucker (1980),, Swamy and Thulasiraman (1981), Tarjan(1983), and Syslo et al.. (1983) Finally, Aho et al. (1974) and Garey and Johnson (1979) have written two of the many books on algorithms from the point of view of the computer scientist.

Applications

Applications of graph theory are mentioned in numerous books covering a wide variety of fields. Some of the more specialized mathematical treatises, which are, either solely or partly, dedicated to graph theory applications are: Busacker and Saaty (1965), Deo (1974), Berman (1978) (a bibliography of applications), Berman and Fryer (1978), Chachra et al. (1979), and Wilson and Beineke (1979) (covering many of the topics of Part II of this book, as well as major applications in communications networks, flowgraphs, and linguistics). The following discuss the application topics covered in Chapter 11 of this book: Harary et al. (1965), Marshall (1971), Roberts (1976, 1978) (social sciences), Deo (1974), Avondo-Bodino (1979) (economics), Cliff et al. (1979) (geography), Earl and March (1979) (architecture), and Tucker (1980) (games).

Operations Research

Marshall (1971), Deo (1974), Christofides (1975), Lawler (1976), Minieka (1978), Avondo-Bodino (1979), Chachra et al. (1979) and Boffey (1982) all contain material directly related to graph theory applications in operations research. Lawler et al. (1985) have compiled a comprehensive treatise on the travelling salesman problem.

Electrical Engineering

Applications of graph theory in this field have been presented by: Stagg and El-Abiad (1968), Balabanian and Bickart (1969), Chen (1971), Deo (1974), Chachra et al. (1979), and Swamy and Thulasiraman (1981).

Industrial Engineering

Applications of graph theoretic concepts to project management, sequencing and line balancing, facilities design, production planning, and control have been presented by Chachra et al. (1979).

Science

Applications of graph theory in science have been reported by Busacker and Saaty (1965) (chemistry, statistical mechanics, and genetics), Marshall (1971) (physics), Temperley (1979) (statistical mechanics), and Rouvray and Balaban (1979) (chemistry).

Civil Engineering

The book by Potts and Oliver (1972) contains an interesting introduction to graph theory and its application in traffic flow prediction. Chachra et al. (1979) introduce graph theoretic models for construction projects, pipeline flows, and transportation problems.

Bibliography

Pages in the text which refer to the following works are given in brackets.

Aho, A.V., Hopcroft, J. and Ullman, J.D. (1974) *The Design and Analysis of Computer Algorithms*, Addison-Wesley, Reading, Mass. [173, 362]

Aho, A.V., Hopcraft, J. and Ullman, J.D. (1983) *Data Structures and Algorithms*, Addison-Wesley, Reading, Mass. [261, 362]

Appel, K.L. and Haken, W. (1976) "Every Planar Map is Four-Colorable", *Bull. Am., Math. Soc., 82*, 711–712. [7]

Assad, A.A. (1978) "Multicommodity Network Flows — A Survey", *Networks, 8*, 37-91. [261]

Avondo-Bodino, G. (1979) "Graph Theory in Operations Research", in *Applications of Graph Theory*, R.J. Wilson and L.W. Beineke, editors, Academic Press, London, Ch.8. [196, 199, 362, 363]

Balabanian, N. and Bickart, T.A. (1969) *Electrical Network Analysis*, Wiley, New York. [363]

Balas, E. (1965) "An Additive Algorithm for Solving Linear Programs with Zero-One Variables", *Operations Research, 13*, 517–546. [294]

Barr, R.S.F., Glover, F. and Klingman, D. (1974) "An Improved Version of the Out-of-Kilter Method and a Comparative Study of Computer Codes", *Math. Prog., 7*, 60–87. [263]

Bazaraa, M.S. and Jarvis, J.J. (1977) *Linear Programming and Network Flows*, Wiley, New York. [362]

Beckman, M., McGuire, C.B. and Winston, C.B. (1956) *Studies in the Economics of Transportation — Part I*, Yale University Press, New Haven. [348]

Behzad, M., Chartrand, G. and Lesniak-Foster, L. (1979) *Graphs and Digraphs*, Wadsworth, Belmont, Calif. [361, 362]

Beineke, L.W. (1967) "The decomposition of complete graphs into planar subgraphs" in *Graph Theory and Theoretical Physics*, Academic Press London, pp 139–154. [72]

Bellman, R. (1958) "On a Routing Problem", *Quart. J. of Appl. Math.*, *16*, 87. [239]

Bellmore, M., Bennington, G. and Lubore, S. (1971) "A Multivehicle Tanker Scheduling Problem", *Transp. Sci.*, *5*, 36–47. [264]

Bennett, G. (1910) "The Eight Queens Problem", *Messenger of Mathematics*, *39*, 19. [216]

Berge, C. (1957) "Two theorems on graph theory", *Proc. Nat. Acad. Sci. (USA) 43*, 842–844. [135]

Berge, C. (1962) *The Theory of Graphs and its Applications*, Methuen, London. [62, 234, 361]

Berge, C. (1973) *Graphs and Hypergraphs* North-Holland, Amsterdam. [361]

Berman, G. (1978) *Applied Graph Theory Bibliography*, Dept. of Combinatorics and Optimization, University of Waterloo, Canada. [362]

Berman, G. and Fryer, K.D. (1978) *Introduction to Applied Graph Theory* René Descartes Foundation, Waterloo, Canada. [362]

Biggs, N.L., Lloyd, E.K. and Wilson, R.J. (1977) *Graph Theory 1736–1936*, Clarendon Press, Oxford. [361]

Boffey, T. B. (1982) *Graph Theory and Operations Research*, Macmillan, London. [363]

Bollobás, B. (1979) *Graph Theory: An Introductory Course*, Springer-Verlag, Berlin. [361]

Bondy, J.A. and Murty, U.S.R. (1976) *Graph Theory with Applications*, Macmillan, London. [228, 230, 361]

Bott, R. and Mayberry, J.P. (1954) "Matrices and Trees", in *Economic Activity Analysis*, O. Morgenstern, editor, Wiley, New York, 391–400. [202]

Bradley, G., Brown, G. and Graves, G. (1977) "Design and Implementation of Large-Scale Primal Transshipment Algorithms", *Management Science, 24*, 1–34. [263]

Braess, D. (1968) "Über ein Paradoxon der Verkehrsplanung", *Unternehmensforschung, 12*, 258–268. [350]

Bron, J. and Kerbosch, H. (1973) "Algorithm 457 — Finding All Cliques of an Undirected Graph", *Comm. ACM., 16*, 575. [218, 230]

Busacker, R.G. and Saaty, T.L. (1965) *Finite Graphs and Networks: An Introduction with Applications*, McGraw-Hill, New York. [195, 196, 220, 281, 362, 363]

Capobianco, M. and Molluzzo, J.C. (1978) *Examples and Couterexamples in Graph Theory*, North-Holland, Amsterdam. [361]

Carré, B. (1979) *Graphs and Networks*, Clarendon Press, Oxford. [361, 362]

Cayley, A. (1857) "On the theory of the analytical forms called trees", *Philos. Mag., 13*, 19–30. [7, 36]

Cayley, A. (1874) "On the mathematical theory of isomers", *Philos Mag., 67*, 444–446. [36]

Cayley, A. (1875) "Über die analytischen Figuren, welche in der Mathematik Bäume genannt werden und ihre Anwendung auf die Theorie chemischer Verbindungen," *Ber. Deut. Chem. Ges., 8*, 1056–1059. [328]

Cayley, A. (1889) "A theorem on trees", *Quart. J. Math., 13*, 26–28. [36]

Cayley, A. (1895) "The theory of groups and graphical representation", *Mathematical Papers 10*, 26–28, Cambridge University Press. [36]

Cayley, A. (1896) "On the analytical forms called trees", *Mathematical Papers 11*, 365–367, Cambridge University Press. [36]

Chachra, V., Ghare, P.M. and Moore, J.M. (1979) *Applications of Graph Theory Algorithms*, North Holland, Amsterdam. [292, 343, 362, 363]

Chen, W.K. (1971) *Applied Graph Theory*, North-Holland, Amsterdam. [363]

Cheston, G.A. (1976) "Incremental algorithms in graph theory", Dept, of Computer Science, Technical Report 91, University of Toronto, Canada. [313]

Cheston, C.A. and Corneil, D.G. (1982) "Graph property update algorithms and their application to distance matrices", *INFOR, 20*, 178–201. [313]

Christofides, N. (1975) *Graph Theory: An Algorithmic Approach*, Academic Press, New York, 62. [176, 229, 294, 362, 363]

Christofides, N. (1976) "Worst Case Analysis of a New Heuristic for the Travelling Salesman Problem", *Management Science Research Report No. 388*, Carnegie-Mellon University, Pittsburgh, USA. [232]

Chvátal, V. (1983) *Linear Programming*, Freeman, San Francisco. [261, 263]

Clarke, S. and Surkis, J. (1968) "An Operations Research Approach to Racial Desegregation of School Systems", *Socio-Econ. Plan. Sci. 1*, 259–272. [265]

Cliff, A.D., Haggett, P. and Ord, J.K. (1979) "Graph Theory and Geography", in *Applications of Graph Theory*, R.J. Wilson and L.W. Beineke, editors, Academic Press, London, Ch. 10. [196, 203, 362]

Corneil, D.G. (1970) "Graph Isomorphism", *Tech. Rep. No. 18*, Dept. of Computer Science, University of Toronto, Canada. [173]

Corneil, D.G. and Graham, B. (1973) "An algorithm for determining the chromatic number of a graph", *SIAM J. Comput. 2*, 311–318. [177]

Couper, A.S. (1858) "Sur une nouvelle théorie chimique", *Ann. Chim. 53*, 469–489. [324]

Crum Brown, A. (1884) "On the theory of isomeric compounds", *Trans. Roy. Soc. Edinburgh, 23*, 707–719. [7, 324]

Cundy, H.M. and Rollett, A.P. (1952) *Mathematical Models*, Clarendon Press, Oxford. [297]

Dantzig, G.B. (1963) *Linear Programming and Extensions*, Princeton University Press. [261, 263]

Dantzig, G.B. (1967) "All shortest routes in a graph", in *Theory of Graphs*, Proc. Int. Symp. Rome, July 1966, Gordon and Breach, New York. [239, 312]

Dayhoff, M.O. and Eck, R.V., editors, (1972) *Atlas of protein sequence and structure, 5*, National Biomedical Research Foundation, Washington DC. [331, 338]

Dempster, M.A.H. (1971) "Two Algorithms for the Timetable Problem", in *Combinatorial Mathematics and its Applications*, D.J.A. Welsh, editor, Academic Press, London. [228]

Deo, N. (1974) *Graph Theory with Applications to Engineering and Computer Science*, Prentice-Hall, Englewood Cliffs, N.J. [196, 361, 362, 363]

de Werra, D. (1970) "On some Combinatorial Problems arising in Scheduling", *INFOR, 8*, 165–175. [228]

Dijkstra, E. (1959) "A Note on Two Problems in Connection with Graphs", *Numerische Mathematik, 1*, 269–271. [238, 239]

Eades, P., Foulds, L.R. and Giffin, J.W. (1982) "An Efficient Heuristic for Identifying a Maximum Weight Planar Subgraph", in *Combinatorial Mathematics IX*, Lecture Notes in Mathematics No. 952, Springer-Verlag, Berlin. [304, 306]

Earl, C.F. and March, L.J. (1979) "Architectural Applications of Graph Theory", in *Applications of Graph Theory*, R.J. Wilson and L.W. Beineke, editors, Academic Press, London, Ch.11. [196, 208, 362]

Edmonds, J. (1965) "Paths, Trees and Flowers", *Canad. J. Math., 17*, 449–467. [179]

Elmaghraby, S.E. (1970) *Some Network Models in Management Science*, Springer-Verlag, New York. [265]

Euler, L. (1736) "Solutio problematis ad geometriam situs pertinentis", *Comment. Academiae Sci. I. Petropolitanae, 8*, 128-140. [3, 9]

Even, S. and Kariv, O. (1975) "An $O(n^{5/2})$ Algorithm for Maximum Matching in General Graphs", *Proc. 16th Annual Symp. on Foundations of Comp. Science, IEEE* 100–112. [180]

Even, S. and Tarjan, R.E. (1975) Network Flow and Testing Graph Connectivity", *SIAM J. Comput. 4*, 507–518. [180]

Fáry, I. (1948) "On straight line representation of planar graphs", *Acta Sci. Math. Szeged 11*, 229–233. [56]

Fisher, M. and Essam, J. (1961) "Some Cluster Size and Percolation Problems", *J. Math. Phys., 2*, 609–619. [323]

Fisher, M.E. and Temperley, H.N.V. (1961) "Dimer Problem in Statistical Mechanics. — An Exact Result", *Philos. Mag., 6*, 1061–1063. [324]

Floyd, R.W. (1962) "Algorithm 97 — Shortest Path", *Comm. ACM, 5*, 345. [239]

Ford, L.R. (1956) "Network flow theory", *Rand Corporation Report, No. 923*, Santa Monica, Calif., USA. [239]

Foulds, L.R. and Robinson, D.F. (1976) "A Strategy for Solving the Plant Layout Problem", *Operational Research Quarterly, 27*, 845–855. [296]

Foulds L.R. and Robinson, D.F. (1978) "Graph Theoretic Heuristics for the Plant Layout Problem", *Int. J. Production Research, 16*, 27–37. [296, 297]

Foulds, L.R., Perara, S.M. and Robinson, D.F. (1978) "Network Layout Procedure for Printed Circuit Design", *Computer Aided Design, 10*, 441–451 [279]

Foulds, L.R. and Robinson, D.F. (1979) "Construction Properties of Combinatorial Deltahedra", *Discrete Applied Mathematics 1*, 75–87. [299, 321]

Foulds, L.R. and Robinson, R.W. (1980) "Determining the Asymptotic Number of Phylogenetic Trees", *Lecture Notes in Mathematics*, No. 829, Combinatorial Mathematics VII, Springer-Verlag, Berlin, 110–126. [37]

Foulds, L.R. and Robinson, R.W. (1981) "Enumeration of Binary Phylogenetic Trees", *Lecture Notes in Mathematics*, No. 884, Combinatorial Mathematics VIII, Springer-Verlag, Berlin, 173–186. [37]

Foulds, L.R. and Robinson, R.W. (1984) "Enumeration of Phylogenetic Trees without Points of Degree Two", *Ars Combinatoria 17A*, 169–183. [37]

Foulds, L.R. (1984) "Maximum Savings in the Steiner Problem in Phylogeny", *J. Theor. Biol., 107*, 471–474. [341]

Foulds, L.R. and Robinson, R.W. (1985) "Counting Certain Classes of Evolutionary Trees with Singleton Labels", *Congressus Numerantium* *44*, 65–88. [37]

Foulds, L.R. (1985) "Traffic Network arc elimination by branch and bound enumeration", *The Arabian Journal for Science and Engineering,* *10*, 149–157. [344]

Foulds, L.R., Gibbons, P.B., and Giffin, J.W. (1985) "Graph Theoretic Heuristics for Facilities Layout: An Experimental Approach", *Operations Research, 33*, 1091–1106. [300]

Foulds, L.R. and Giffin, J.W. (1985) "A graph-theoretic heuristic for minimizing total transport cost in facilities layout", *Int. J. Production Research, 23*, 1247–1257. [300, 309, 315, 318]

Foulds, L.R. and Robinson, R.W. (1988) "Enumerating Phylogenetic Tress with Multiple Labels", *Discrete Mathematics, 72*. [37]

Foulds, L.R. (1989) "The Application of the Theory of Directed Graphs in the Social Sciences", in *Operational Research and the Social Sciences*, M.C. Jackson, P. Keys, and S.A. Cropper eds, Plenum, New York, 183–188. [197]

Foulds, L.R. (1993) "LayoutManager: A Decision Support System for Facilities Planning", *Proceedings of the IFIP TC5/WG5.3/IFAC International Working Conference on Knowledge Based Hybrid Systems, KNOWSHEM '93*, Budapest, Hungary, 20–22 April 1993, North Holland, 293–300 [318]

Frank, M. and Wolfe, P. (1956) "An Algorithm for Quadratic Programming", *Naval Research Logistics Quarterly, 3*, 95–110. [356]

Fratta, L., Gera, M. and Kleinrock, L. (1973) "The Flow Deviation Method: An Approach to Store-and-Forward Communication Network Design", *Networks, 3*, 97–133. [265]

Fulkerson, D.R. (1961) "An Out-of-Kilter Method for Minimal Cost Flow Problems", *J. SIAM, 9*, 18–27. [263]

Gabow, H.N. (1976) "An Efficient Implementation of Edmond's Algorithm for Maximum Matching on Graphs", *J. ACM 23*, 221–234. [180]

Gallai, T. (1959) Über extreme Punkt-und Kantenmengen. *Ann. Univ. Sci. Budapest Eötvös Sect. Math, 2*, 133–138. [128]

Garey, M.R., Graham, R.L. and Johnson, D.S. (1977) "The complexity of computing Steiner minimal trees", *SIAM J. Appl. Math., 32*, 835–859. [156, 157, 159]

Garey, M.R. and Johnson, D.S. (1977) "The Rectilinear Steiner problem is NP-complete", *SIAM J. Appl. Math. 32*, 826–834. [156]

Garey, M.R. and Johnson, D.S. (1979) *Computers and Intractability: A Guide to the Theory of NP-Completeness*, Freeman, San Francisco. [155, 362]

Gibbons, A. (1985) *Algorithmic Graph Theory*, Cambridge University Press Cambridge. [362]

Giffin, J.W. and Foulds, L.R. (1987) "Facilities layout generalized model solved by n-boundary shortest path heuristics", *European J. of Operational Research, 28*, 382–391. [300, 310, 315]

Gilmore, P.C. and Hoffman, A.J. (1964) "A Characterization of Comparability Graphs and of Interval Graphs", *Can. J. Math., 16*, 539–548. [329, 341]

Glaisher, J.W. (1874) "On the Problem of Eight Queens", *The Phil. Mag., Series 4, 48*, 457. [216]

Glover, F. and Klingman, D. (1977) "Network Applications in Industry and Government", *AIIE Transactions, 9*, 383–376. [261]

Glover, F. and Klingman, D. (1978) "Some Recent Practical Misconceptions about the State-of-the-Art of Network Algorithms", *Operations Research, 26*, 370–379. [263]

Golden, B.L. and Magnanti, T.L. (1977) "Deterministic Network Optimization: A Bibliography", *Networks, 7*, 149–183. [261]

Good, I.J. (1946) "Normal recurring decimals", *J. London Math. Soc. 21*, 167–172. [227]

Grunbaum, B. (1967) *Convex Polytopes*, Wiley, New York. [311, 321]

Guan, Mei-gu. (1962) "Graphic Programming using Odd or Even Points", *Chinese Math., 1*, 272-277. [227]

Guthrie, F. (1880) "Note on the colouring of maps", *Proc. Royal Soc. Edinburgh, 10*, 727–788. [7]

Guy, R.K. (1960) "A combinatorial problem", *Bull. Malayan Math. Soc., 7*, 68–72. [72]

Guy, R.K. (1972) "Crossing number of graphs", in *Graph Theory and Applications*, Springer-Verlag, New York, 111–124. [72]

Hanan, M. and Kurtzberg, J.M. (1972) "Placement Techniques", in *Design Automation of Digital Systems*, H. Breuer, editor, Prentice-Hall, Englewood Cliffs, New Jersey. [230]

Harary, F., Norman, R.Z. and Cartwright, D. (1965) *Structural Models: An Introduction to the Theory of Directed Graphs*, Wiley, New York. [196, 362]

Harary, F. (1969) *Graph Theory*, Addison-Wesley, Reading, Mass. [361]

Harary, F. and Palmer, E.M. (1973) *Graphical Enumeration*, Academic Press, New York. [361]

Harary, F., Robinson, R.W. and Schwenk, A.J. (1975) "Twenty step algorithm for determining the asymptotic number of trees of various species", *J. Austral. Math. Soc., 20A*, 483–503. [37]

Heap, B.P. (1966) "The Enumeration of Homeomorphically Irreducible Star Graphs", *J. Math. Phys.*, *7*, 1582–1587. [323]

Heawood, P.J. (1890) "Map Colour Theorems", *Quart. J. Math.*, *24*, 332–338. [7]

Henze, H.R. and Blair, C.M. (1931) "The number of isomeric hydrocarbons of the methane series", *J. Amer. Chem. Soc.*, *53*, 3077–3085. [324, 328]

Higgins, W. (1789) *A Comparative View of the Phylogistic and Anti-Phylogistic Theories*, Murray, London. [324]

Hoffman, E.J., Loessi, J. and Moore, R.C. (1969) "Constructions for the Solution of the m-Queens Problem", *Mathematics Magazine*, *42*, 66–72. [216]

Hopcroft, J.E. and Tarjan, R.E. (1972) "Isomorphism of Planar Graphs", in *Complexity of Computer Computation*, R.E. Miller and J.W. Thatcher, editors, Plenum, New York, 131–152. [173]

Hopcroft, J. and Tarjan, R.E. (1974) "Efficient Planarity Testing", *J. ACM* *21*, 549–568. [70, 152, 168, 295]

Howard, R.A. (1971) *Dynamic Probabilistic Systems, I: Markov Modes*, Wiley, New York. [233]

Jensen, P.A. and Barnes, J.W. (1980) *Network Flow Programming*, Wiley, New York. [362]

Kasteleyn, P.W. (1961) "The Statistics of Dimers on a Lattice", *Physica*, *27*, 1209–1225. [324]

Kaufmann, A. (1967) *Graphs, dynamic programming and finite games*, Academic Press, New York. [227, 234, 361]

Kaufman, A. (1972) *Points and Arrows*, Transworld Publishers, London. [362]

Kempe, A.B. (1879) "On the geographical problem of four colours", *Am. J. Math.*, *2*, 193–204. [7]

Kennington, J.L. and Helgason, R.V. (1980) *Algorithms for Network Programming*, Wiley, New York. [261, 263, 264, 362]

Kevin, V. and Whitney, M. (1972) "Algorithm 442 — Minimum Spanning Tree", *Comm. ACM*, *15*, 273. [238]

Kirchhoff, G. (1847) "Uber die Auf lösung der Gleichungen, auf welche man bei der Untersuchung der linearen Verteilung galvanischer Ströme geführt wird", *Ann. Phys. Chem.*, *72*, 497–508. [8, 270]

Kirkby, M.J. (1976) "Tests on a random network model and its application to basin hydrology", *Earth Surface Processes*, *1*, 197–212. [207]

Knodel, W. (1969) *Graphtheoretische Methoden und ihre Anwedungen*, Springer-Verlag, Berlin, 56–59. [350]

König, D. (1931) "Graphen und Matrizen", *Mat. Fiz. Lapok.*, *38*, 116–119. [129, 135]

König, D. (1936) *Theorie der endlichen und unendlichen Graphen*, Akademische Verlagsgesellschaft, Leipzig. Reprinted: Chelsea, New York, 1950. [361]

Kraitchik, D. (1960) *Mathematical Recreations*, 2nd. edition, New York, 238. [217]

Krejcirik, M. (1969) "Computer Aided Plant Layout", *Computer Aided Design, 2*, 7–17. [296]

Kruskal, J.B. (1956) "On the Shortest Spanning Subtree of a Graph and the Travelling Salesman problem", *Proc. Am. Math. Soc., 7*, 48–50. [234]

Kuhn, H.W. (1955) "The Hungarian Method for the Assignment Problem", *Naval Research Logistics Quarterly, 2*, 83–97. [229]

Kuratowski, K. (1930) "Sur le probléme des courbes gauches en topologie", *Fund. Math. 15*, 271–283. [62]

Laurent, A. (1864) "Recherches sur les combinaisons azotées", *Ann. Chim. Phys., 18*, 266–298. [324]

Lawler, E.L. and Wood, D.E. (1966) "Branch and Bound Methods: A Survey", *Operations Research, 14*, 699–719. [294]

Lawler, E.L. (1976) *Combinatorial Optimization : Networks and Matroids*, Holt, Reinhart and Winston, New York. [191, 362, 363]

Lawler, E.L., Lenstra, J.K., Rinnooy-Kan, A. and Shmoys, D.B. (1985) *The Travelling Salesman Problem*, Wiley-Interscience, New York. [232, 363]

Le Blanc, L.J. (1975) "An Algorithm for the Discrete Network Design Problem", *Transportation Science, 9*, 183–199. [349, 350, 354, 355]

Lin, S. (1965) "Computer Solutions of the Travelling Salesman Problem", *Bell Syst. Tech. J., 44*, 2245–2269. [232]

Liu, C.L. (1968) *Introduction to Combinatorial Mathematics*, McGraw-Hill, New York. [142]

Lovasz, L. (1975) "Three short proofs in graph theory", *J. Combinatorial Theory 19B*, 269–271. [136]

Mandl, C. (1979) *Applied Network Optimization*, Academic Press, New York. [362]

Marshall, C.W. (1971) *Applied Graph Theory*, Wiley-Interscience, New York. [195, 362, 363]

Matula, D.W., Marble, G. and Isaacson, J.D. (1972) "Graph colouring algorithms", in *Graph Theory and Computing*, R.C. Read, editor, Academic Press, New York, p. 109. [176, 232]

McDiarmid, C. (1976) "Determining the chromatic number of a graph", Report STAN-CS-76-576, Dept. of Computer Science, Stanford University, USA. [177]

Metra Group (1964) Sema, Le Modele ATCODE, France. [351]

Meyniel, M. (1973) "Une condition suffisante d'existence d'un circuit Hamiltonien dans un graph oriente", J. Combinatorial Theory, 14B, 137–147. [100]

Minieka, E. (1978) Optimization Algorithms for Networks and Graphs, Marcel Dekker, New York. [294, 362, 363]

Moon, J.W. (1967) "Various Proofs of Cayley's Formula for Counting Trees", in A Seminar on Graph Theroy, F. Harary, editor, Holt, Reinhart and Winston, New York, pp. 70–78. [36]

Moon, J.W. (1970) "Counting Labelled Trees", Canad. Math. Congress 1970, Montreal, William Clowes and Sons, London. [361]

Moore, E.F. (1957) "The Shortest Path Through a Maze", Proc. Int. Symp. on The Theory of Switching, III, 285. [239]

Moore, J.M. (1976) "Facilites Design with Graph Theory and Strings", Omega, 4, 193–203. [296]

Murchland, J.D. (1970) "Braess's Paradox of Traffic Flow", Transportation Research, 4, 391–394. [350]

Nicholson, T.A.J. (1968) "Permutation procedure for minimizing the number of crossings in a network", Proc. Inst. Elec. Eng., 115, 21–26. [281]

Nijenhuis, A. and Wilf H. (1975) Combinatorial Algorithms, Academic Press, New York. [227]

Owens, A. (1971) "On the biplanar crossing number", IEEE Trans. on Circuit Theory, 18, 277–280. [281]

Panagiotakopoulos, D. (1976) "Multicommodity Multi-transformed Network Flows with Application to Residuals Management", Management Science, 22, 874–882. [265]

Penny, E.D., Foulds, L.R. and Hendy, M.D. (1982) "Testing the Theory of Evolution by Comparing Phylogenetic Trees Constructed from Five Different Protein Sequences", Nature, 297, 197–200. [329, 338]

Phillips, D.T. and Garcia-Diaz, A. (1981) Fundamentals of Network Analysis, Prentice-Hall, Englewood Cliffs, N.J. [362]

Plane, D.R. and McMillan, C. (1971) Discrete Optimization, Prentice-Hall, Englewood Cliffs, N.J. [362]

Polya, G. (1937) "Kombinatorische Anzahlbestimmungen für Gruppen, Graphen und chemische Verbindungen", Acta Math., 68, 145–254. [36, 325]

Polya, G. (1940) "Sur les types des propositions composées", *J.Symb. Logic, 5*, 98–103. [36, 325]

Potts, R.B. and Oliver, R.M. (1972) *Flows in Transportation Networks*, Academic Press, New York. [351, 363]

Prim, R.C. (1957) "Shortest Connection Networks and some Generalizations", *Bell Syst. Tech. J., 36*, 1389. [234]

Rao, M.R. and Zionts, S. (1968) "Allocation of Transportation Units to Alternative Trips", *Operations Research, 16*, 52–63. [264]

Read, R.C. (1969) "Teaching graph theory to a computer", in *Recent Progress in Combinatorics*, W.T. Tutte, editor, Academic Press, New York, 161–173. [177]

Read, R.C. and Harary, F. (1970) "The enumeration of tree-like polyhexes", *Proc. Edinburgh Math. Soc.* (2) *17*, 1–13. [325]

Read, R.C. (1979) "Algorithms in Graph Theory", in *Applications of Graph Theory*, R.J. Wilson and L.W. Beineke, editors, Academic Press, London, Ch. 13. [173, 362]

Recski, A. (1989) *Matroid Theory and its Application in Electric Network Theory and Statics*, Springer Verlag, Berlin. [191, 362]

Redfield, J.H. (1927) "The theory of group-reduced distribution", *Am. J. Math., 49*, 433–455. [36]

Robbins, H.E. (1939) "A theorem on graphs with an application to a problem in traffic control", *Am. Math. Mthly, 46*, 281–283. [347]

Roberts, F.S. (1976) *Discrete Mathematics Models, with Applications to Social, Biological, and Environmental Problems*, Prentice-Hall, Englewood Cliffs, New Jersey. [347, 362]

Roberts, F.S. (1978) "Graph Theory and its Application to Problems of Society", *CBMS-NSF Reg. Conf. Series in Appl. Math.*, SIAM. [195, 362]

Robinson, R.W. (1975) "Counting Arrangements of Bishops", Springer Lecture Notes in Mathematics, *Combinatorial Mathematics IV, 560*, 198. [217]

Robinson, D.F. and Foulds, L.R. (1980) *Digraphs: Theory and Techniques*, Gordon and Breach, London. [108, 220, 291, 362]

Rosenblatt, D. (1957) "On the Graphs and Asymptotic Forms of Finite Boolean Relation Matrices and Stochastic Matrices", *Naval Research Logistics Quarterly, 4*, 151–161. [233]

Rouvray, D.H. and Balaban, A.T. (1979) "Chemical Applications of Graph Theory", in *Applications of Graph Theory*, R.J. Wilson and L.W. Beineke, editors, Academic Press, London, Ch.7. [328, 363]

Shreve, R.L. (1966) "Statistical law of stream numbers", *J.Geology, 74*, 17–37. [203]

Skupien, Z. (1966) "Locally Hamiltonian and planar graphs", *Fund. Math.*, *58*, 193–200. [306]

Smart, J.S. (1967) "A comment on Horton's law of stream numbers", *Water Resources*, *4*, 1001–1014. [203]

Smart, J.S. (1969) "Topological properties of channel networks", *Bull. Geolog. Soc. Am.*, *80*, 1757–1774. [203]

Smith, D.K. (1982) *Network Optimisation Practice*, Ellis Horword, Chichester, UK. [362]

Stagg, G.W. and El-Abiad, A.H. (1968) *Computer Systems in Power Systems Analysis*, McGraw-Hill, London. [363]

Strahler, A.N. (1964) "Quantitative geomorphology of drainage basins and channel networks", in *Handbook of Applied Hydrology*, V.T. Chow editor, McGraw-Hill, New York, 4.40–4.74. [203]

Sussenguth, E.H. (1965) "A Graph Theoretic Algorithm for Matching Chemical Structures", *J. Chem. Doc.*, *5*, 36–43. [173, 328]

Swamy, M.N.S. and Thulasiraman, K. (1981) *Graphs, Networks and Algorithms*, Wiley Interscience, New York. [362, 363]

Sylvester, J.J. (1909) *Collected Math. Papers of James Joseph Sylvester*, *Vol. III*, Cambridge University Press, Cambridge, 148–206. [324]

Syslo, M.M., Deo, N. and Kowalik, J.S. (1984) *Discrete Optimization Algorithms*, Prentice-Hall, New Jersey. [362]

Takahashi, H. and Matsuyama, A. (1980) "An Approximate Solution for the Steiner Problem in Graphs", *Math. Japonica*, *6*, 573–577. [233]

Tarjan, R.E. (1983) *Data Structures and Network Algorithms*, CBMS44, SIAM, Philadelphia. [261, 362]

Temperley, H.N.V. (1979) "Graph Theory and Discrete Statistical Mechanics", in *Application of Graph Theory*, R.J. Wilson and L.W. Beineke, editors, Academic Press, London, Ch.6. [323, 363]

Tinkler, K.J. (1973) "The topology of rural periodic market systems", *Geografiska Annaler*, *55B*, 121–133. [202]

Tomassen, C. (1980) "Planarity and duality of finite graphs" *J. Comb. Th.*, *29B*, 244–271. [304]

Tran, H.V. (1982) *"Layout of the Proposed University of Canterbury Law Library"*, Unpublished MSc thesis, Department of Economics, University of Canterbury, Private Bag 4800, Christchurch, New Zealand. [302]

Trudeau, R.J. (1976) *Dots and Lines*, Kent State University Press, Kent, Ohio, USA. [361]

Tucker, A. (1980) *Applied Combinatorics*, Wiley, New York. [196, 361, 362]

Turan, P. (1977) "A Note of Welcome", *J. Graph Theory*, *1*, 7–9. [71]

Tutte, W.T. (1947) "The factorizations of linear graphs", *J. London Math. Soc.*, *22*, 107–111. [136]

Tutte, W.T. (1962) "A Census of Planar Triangulations", *Canad. J. Math.*, *14*, 21–38. [208, 209, 223]

Tutte, W.T. (1984) *Graph Theory*, Addison-Wesley, Reading, Mass. [361]

von Randow, R. (1975) *Introduction to the Theory of Matroids*, Springer Lecture Notes in Mathematical Economics, *No. 109*. [191, 362]

Wang, C.C. (1974) "An algorithm for the chromatic number of a graph", *J. ACM. 21*, 385–391. [177]

Wardrop, J.G. (1952) "Some Theoretical Aspects of Road Traffic Research" *Proceedings of Institute of Civil Engineers, 2*, 325–378. [347]

Welsh, D.J.A. and Powell, M.B. (1967) "An upper bound on the chromatic number of a graph and its application to timetabling problems". *The Computer J., 10*, 85. [176]

Welsh, D.J.A. (1976) *Matroid Theory*, Academic Press, New York. [191, 362]

Werrity, A. (1972) "The topology of stream networks", in *Spatial Analysis in Geomorphology*, R.J. Chorley, editor, Methuen, London, 167–196. [207]

White, N. (1986) Editor, *Theory of Matroids*, Encyclopedia of mathematics and its applications, vol. 26, Cambridge University Press, Cambridge, U.K. [191, 362]

White, N. (1992) Editor, *Matroid Applications*, Encyclopedia of mathematics and its applications, vol. 40, Cambridge University Press, Cambridge, U.K. [191, 362]

White, W.W. and Bomberault, A.M. (1969) "The Network Algorithm for Empty Freight Car Allocation", *IBM Systems J., 9*, 147–169. [264]

Whitney, H. (1935) "On the Abstract Properties of Linear Dependence", *Am. J. Math., 57*, 509–533. [183]

Williams, M.R. (1968) "A graph theory model for the solutions of timetables", Ph. D. Thesis, University of Glasgow, Scotland. [176, 232]

Wilson, R.J. (1972) *Introduction to Graph Theory*, Oliver and Boyd, London. [361]

Wilson, R.J. and Beineke, L.W. (1979) *Applications of Graph Theory*, Academic Press, London. [362]

Winston, W.L. (1991) *Operations Research: Applications and Algorithms*, Second Edition, PWS-Kent, Boston. [294]

Witzgall, D. and Zahn, C.T. (1965) "Modifications of Edmond's Algorithm for Maximum Matching of Graphs", *J. Res. Nat. Bur. Std., 69B*, 91–98. [180]

Yaged, B. (1973) "Minimum Cost Routing for Dynamic Network Models",
    Networks, 3, 193–224. [265]

Yaglom, A. and Yaglom, I. (1964) *Challenging Mathematical Problems with
    Elementary Solutions*, Holden-Day, San Francisco, 78. [216]

Zuckerkandl, E. and Pauling, L. (1965) "Molecules as Documents of Evo-
    lutionary History", *J. Theor. Biol., 8*, 357–366. [331]

Index

1-factor, 136
1-isomorphic, 18
2-isomorphic, 18

Adjacent, 9
Algebraic sum, 270
Algorithm, 145, 146, 294, 362
  efficiency, 150
  graph analysis, 161
  graph optimization, 174–180
  greedy, 188, 190, 308
  input, 146
  order, 149
  output, 160
Aminoacetone, 326
Arc, 11
  converse, 94, 107
  elimination, 350
  in kilter, 263
  return, 247
Arborescence, 100, 202
Architecture, 196, 207–210, 362
Assignment problem, 229
Automorphism, 13

Bifurcation ratio, 206
Biology, 328
Block, 18
Block plan, 293, 296, 218
Blossom, 179
Borrow pits, 343
Branch, 9, 27
Braess's paradox, 350
Breakthough, 252
Bridge, 18, 181, 347

Centres, 31, 125
  absolute p-, 229
  location of, 229
Chemistry, 7, 324
Chinese Postman's problem, 227
Chord, 38, 101
Chromatic number, 132
Chromatic partitioning problem, 133
Chromatic polynomial, 177
Civil engineering, 343
Circuit, 8
  fundamental, 8

Clique, 215–217
Closeness ratings, 292
Closure, 48
Coalescement, 165, 334
Colour class, 132
Colouring, 124, 132–134
    algorithms, 175–179
    c-, 132
    c-colourable, 132
    sequential, 176
Complement, 23, 215
Complexity, 149–160
    expected time, 149
    time, 149
    worst-case, 149
Component, 18, 238
    size sequence, 110
    strong, 97
    unilateral, 97
    weak, 97
Computer science, 196, 362
Condensation, 98
Connectivity, 17–26, 93–98, 164, 347
    algorithm, 164–168
    edge, 25
Connector problem, 228
Consistency of choice, 198
Consumption matrix, 201
Continuum statistical mechanics, 340
Contraction, 304
Converse digraph, 94
Covering, 123–127
    dimer, 136
    edge, 124
        number, 124
    vertex, 124

number, 125
Crossing number, 71
Cut, 248
    Capacity, 248
    Minimal, 248
Cut-set, 67, 98
    directed, 104
    fundamental, 104, 272
    matrix, 84–85
        fundamental, 84–85, 104
Cut-vertex, 18
Cycle, 18, 93, 185
    even, 18
    fundamental, 38, 272
        algorithm, 173
    Hamiltonian, 47, 188, 232
    matrix, 79
        fundamental, 81–84
    matroid, 184
    odd, 18
Cyclomatic number, 65
Cytochrome c, 330, 338

Darwin, C., 329
Deficiency, 140
Degree, 18
Deltahedron
    heuristic, 296–302
    n-boundary, 310
    super-, 316
Diameter, 31
Digraph, 11, 233, 362
    acyclic, 94
    cyclic, 94
    disconnected, 94
    Eulerian, 98

Hamiltonian, 99
  strong, 94
  unicursal, 98
  unilateral, 94
  weak, 94
  weighted, 11
Dimer covering, 136, 324
Directed graph, (**see** Digraph)
Distance, 31
Dodecahedron, 4
Dominance, 123, 130
  edge, 130
    number, 130
  vertex, 130
    number, 130
Dominating set, 130
Dual, 64–67
  geometric, 64–67
  self-, 70

Eccentricity, 31
Economics, 196, 199–203, 362
Edge, 9
  boundary, 57
  contraction, 177
  orientation, 163
  replacement, 297
Electrical engineering, 269, 363
Enumeration
  graph, 361
  isomer, 327
  phylogeny, 35–38
  physics, 323
  tree, 7, 35, 361
Euler, L., 3
  formula, 57

Euler trail, 43, 188, 227
Exact cover of 3-sets, 155

Face, 56
Facilities layout, 292–319
Fibrinopeptide, 338
Forest, 30
  spanning, 38
Four colour theorem, 6

Gale optimal, 189
Game theory, 196, 210, 233, 362
  digraph, 233
    vertex
      closing, 233
      kernel, 234
      starting, 233
    players, 233
Genetics, 328
Geography, 196, 203–207, 362
Graph, 7, **9**
  acyclic, 18
  bipartite, 23
  complete, 18
  connected, 18
  disconnected, 18
  embedded, 55
  Eulerian, 43–46
  Hamiltonian, 46–50
  homeomorphic, 60
  interval, 328
  invariant, 173
  k-connected, 18
  Kuratowski, 55
  mixed, 359
  n-partite, 23

nonplanar, 55
nonseparable, 18
null, 21
oriented, 12, 347
planar, 55
    maximally, 59
plane, 55
random, 181
regular, 21
self-complementary, 23
self-dual, 70
separable, 18
thickness, 70
unicursal, 44
weighted, 11
Graphic notation, 7
Greedy, 174
    algorithm, 174–175, 188, 308

Haemoglobin, 338
Hamilton, W.R., 4
Hamiltonian
    cycle, 47, 188, 232
    graph, 46–50
    path, 50
Heuristics, 231, 231–233
Hungarian method, 229

Incident, 9
Indegree, 19, 114, 115
Independence, 123, 127–129
    edge, 127
        number, 127
    vertex, 127
        number, 127
Independent set, 127
Industrial engineering, 291, 363

Isomer, 7
    enumeration, 327
Isomorphic, 13
Isomorphism, 13
    detection of, 170
    1-isomorphic, 18
    2-isomorphic, 18

Join, 9, 17 ·
Junction, 9

Kaliningrad, 45
Kirchhoff's laws, 270–279
    current postulate, 270
    voltage postulate, 270
Königsberg bridge problem, 3, 43, 45, 210
Kuratowski, K., 55
    graphs, 55

Labelling method, 250–256
Leontif economic model, 199
Lexicographically minimum, 188
Lexicographically smaller, 188
Line, 9
Link, 9
Logical numbering, 114

Matching, 124, 134–142
    algorithms, 179–180
    augmenting, 135
    complete, 137
    matroid, 185
    perfect, 136
    unaugmentable, 135
Matrix, 75–92
    digraph,
        adjacency, 77

cut-set, 84–87
  fundamental semicycle, 101
  incidence, 78
  semicycle, 100
  semipath, 106
graph, 75–92
  adjacency, 76–77, 147
  cycle, 75, 80–84
    fundamental, 81–84
  cut-set, 75, 84–87
    fundamental, 85–87
  incidence, 77–79, 147
    reduced, 78, 88
  path, 75, 90–91
Matroid, 183–191
  base, 184
  bipartite, 186
  co-base, 187
  co-cycle, 187
  co-graphic, 186
  cut-set, 184
  cycle, 184
  dependent, 185
  duality, 186
  elements, 184
  Eulerian, 186
  graphic, 186
  independent set, 184
  isomorphic, 186
  k-uniform, 186
  matching, 185
  planar, 186
  rank, 185
Maximum flow problem, 247
Maximum parsimony, 330
Metric, 31
Mine ventilator location, 232

Minimum cost flow problem, 256–261
Molecule representation, 325
Multicommodity network problem, 263
  linear model, 264
Multigraph, 10
  directed, 11
  Eulerian, 44
  unicursal, 44

n-boundary model, 310
n-cycle, 18
Network, 11, 362
  communications, 230, 2365
  electrical, 230, 270, 289
  facility location on a, 230
  node, 11
  sink, 11
  source, 11
  transportation, 246–265
Network flow
  backward, 250
  conservation of, 247
  forward, 250
Node, 11
Nondeterministic polynomial, 151
NP-complete, 151, 155
NP-hard, 155
Nucleotide, 330

Operations research, 225, 363
Outdegree, 18, 115

Paraffins, 326
Path, 17, 93
  alternating, 135
  Hamiltonian, 50
  matrix, 90–91

Percolation process, 323
Pfaffian, 324
Phylogeny, 37
  magnitude of, 37
  order of, 37
  planted, 37
Physical systems, 323, 363
Physics, 8, 323, 363
Planarity
  Detection of, 60, 168–169
Point, 9
Polynomial transformation, 154
Predecessor, 11
Principle of directional duality, 107
Printed circuit board, 279, 280
Production planning, 291
Project selection, 230
Pseudodigraph, 11
Pseudograph, 11
  self-dual, 70
Puzzles, 196, 210–218
  eight queens, 216
  knight's tour, 216
  n-queens, 216

Radius, 31
Ramsey number, 215
Rank, 64, 103
Reachable, 94
Region, 56
  exterior, 57
  interior, 57
Relationship chart, 292

Science, 161, 363
Search, 161

breadth-first, 161
depth-first, 162–164
Semicycle, 94
  fundamental, 101
Semipath, 94
Semitrail, 94
Semiwalk, 94
Shortest path problems, 228,
  239–246
Sink, 11
Social hierarchy, 197
Social sciences, 195, 197–199, 362
Source, 11
Social status, 197
Star, 125
Steiner problem in graphs, 233
Steiner problem in phylogeny,
  155, 332
Steiner minimal tree, 156
Steiner vertex, 158, 335
Storage problem, 232
Stream magnitude, 207
Stream order, 205
Subdigraph, 97
Subgraph, 13
  c-, 176
  induced, 13
  spanning, 13
Substitution, 332
Subtournament, 110
Successor, 11
Supergraph, 13

Thickness, 70
Timetabling problem, 227
Tournament, 108–120

acyclic, 114–115
strong, 111–114
wins analysis, 118–120
Traffic network, 344
Trail, 17, 93
Euler, 43
open, 44
Transportation problem, 344
Transshipment
capacitated model, 229, 261–263
Travelling salesman problem, 228, 231
Tree, 7, **27**
arborescence, 100
binary, 32
directed, 100
enumeration, 35
free, 32
palm, 163
phylogenetic, 35, 329
rooted, 32
search, 162
spanning, 37, 157, 184, 188, 331
algorithm, 169–171
minimal, 156
Triangle, 18
Triangulation, 59
Turing machine, 152
deterministic, 153
nondeterministic, 154

Unicursal, 44, 90, 98
Utilities, 53

Valence, 18
Vertex, 9
central, 31
connectivity, 17
current, 170
degree, 18
distance, 31
indegree, 19
internal, 21
isolated, 21
matched, 179
outdegree, 18
pendant, 21
pseudo-, 179
relocation, 299
unmatched, 179
valence, 18

Walk, 17, 93
closed, 17, 93
open, 17
spanning, 93
unicursal, 90
Wheel, 304
expansion, 305
hub, 304
rim, 304

Universitext   *(continued)*

Rotman: Galois Theory
Rubel/Colliander: Entire and Meromorphic Functions
Sagan: Space-Filling Curves
Samelson: Notes on Lie Algebras
Schiff: Normal Families
Shapiro: Composition Operators and Classical Function Theory
Simonnet: Measures and Probability
Smith: Power Series From a Computational Point of View
Smoryński: Self-Reference and Modal Logic
Stillwell: Geometry of Surfaces
Stroock: An Introduction to the Theory of Large Deviations
Sunder: An Invitation to von Neumann Algebras
Tondeur: Foliations on Riemannian Manifolds
Zong: Strange Phenomena in Convex and Discrete Geometry